MY YEAR WITH ELEANOR

My Year with Eleanor

A MEMOIR

Noelle Hancock

An Imprint of HarperCollinsPublishers

HarperCollins books may be purchased for educational, business, or sales promotional use. For information please write: Special Markets Department, HarperCollins Publishers, 10 East 53rd Street, New York, NY 10022.

A hardcover edition of this book was published in 2011 by Ecco, an imprint of HarperCollins Publishers.

FIRST ECCO PAPERBACK EDITION PUBLISHED 2012.

Library of Congress Cataloging-in-Publication Data has been applied for.

ISBN 978-0-06-187501-4

13 14 15 16   OV/RRD   10 9 8 7 6 5 4 3 2

For "Matt,"
who was my rock throughout
this entire process,

For my loving parents,
who showed me the importance of
staying down-to-earth,

And for Eleanor,
who taught me to fly

Chapter One

Your life is your own. You mold it. You make it.
All anyone can do is to point out ways and means
which have been helpful to others. Perhaps they
will serve as suggestions to stimulate your own
thinking until you know what it is that will fulfill
you, will help you to find out what you want to do
with your life.

—ELEANOR ROOSEVELT

I was lying on a beach in Aruba, mulling a third piña colada, when I received a phone call announcing I'd been laid off from my job. The call came, ironically, on my company cell phone. I'd brought it with me to the beach in case something came up at work.

Something came up.

"They're shutting us down!" squeaked my coworker Lorena.

"W-what?"

"The whole website has been closed down." She sounded like she'd been crying. "We're all out of a job."

I sprang forward on my lounge chair and struggled to free my butt, which had sunk between the vinyl straps. "What are you talking about?" I shook my head in disbelief.

"They called us into a meeting and announced it this afternoon. It took everyone by surprise."

"Why didn't anyone call me?"

"They've been trying, but the office has some kind of block on international calls. I'm calling you from my cell," she said, dropping into a low, conspiratorial whisper. "I thought you'd rather hear it from a friend first."

"But this doesn't make any sense. We're doing so well!" Our online readership had been steadily climbing. Just last week, our website had drawn a million page views in one day.

"Something about cutting costs." Her voice was a little loose. I listened closely and heard loud conversations and Bon Jovi in the background.

"Are you at a bar?" I asked, confused.

"Yeah, the whole staff is at that Irish pub across the street from the office. Listen, I have to get back. I'll call you later, okay?"

When I hung up the phone, I saw my freshly tanned fingers tremble slightly. I stared straight ahead without really seeing anything.

"Who was that?" Matt asked from the lounge chair next to me.

"That was the office," I said dully. "I've been laid off."

"Wait—*what*?" Matt threw down his newspaper. He swung his legs around so he was facing me.

"They've shut down the entire company," I continued in that odd emotionless voice. "Announced in a meeting this afternoon."

"Oh, baby, I'm so sorry. Is there anything I can do?"

He grabbed my hand and I felt the faint squish of sunscreen. Still, I couldn't bring myself to meet his gaze. I was stuck in one of those trances where it appears some invisible hand has smeared itself over your world. And, in a way, it had. It could've been an impressionist painting: *Girl Without a Job Sitting by the Sea,* oil on canvas, 2008.

A ringing sound jerked me out of my daze. I turned and watched Matt grope inside our beach tote for his cell phone. As a political reporter for the most highly regarded newspaper in the country, Matt was also accustomed to answering work calls while on vacation. Just as he found it, the ringing stopped and a chime sounded signaling he had a voice mail.

He peered at the caller ID screen under the glare of sunlight. "Crap, it's work. My editor probably wants me to make some calls for that story that's running tomorrow." He ran an anxious hand through his thick brown hair.

"I'll be fine. Go call him back. I need a moment alone to process this anyway."

"Don't be ridiculous. I'm not leaving you like this."

"Like what?" I said, forcing what I hoped was a convincing smile. "Sitting in a tropical paradise? Seriously, go make your call."

Matt scurried off toward our hotel room, casting a few worried glances over his shoulder. When he disappeared around the corner, I let my smile fade. I felt as though I'd been riding in a car and the driver had unexpectedly slammed on the brakes. Everything had stopped. I was shocked and confused, but also embarrassed for the person I was a few minutes ago who didn't see this coming.

My eyes drifted to the stack of celebrity magazines next to my chair. The one on top was splayed open, Aruba's aggressive trade winds flipping its pages, creating a mini moving picture, the famous Jessicas, Jennifers, and Kates of the world morphing into one another, much the way they do in real life. I'd been reading the magazines for work. For the last several years, I'd worked as a pop culture blogger, churning out stories on a half-hourly basis. In turn, celebrities provided me with constant material by getting married, getting divorced, getting arrested, getting too fat, getting too thin, or just leaving the house for coffee. Yes, the job was fairly absurd, but at nearly six figures, so was the salary.

Twenty feet away, palm trees waved fiercely. We'd been told not to put our chairs under them because coconuts can drop and bonk people on the head, knocking them unconscious. I had a sudden urge to move my chair over there. Instead, I stood up and crunched through the sand toward the hotel. I marched down the steps of the hotel pool and plowed through the shallow end, bouncing from leg to leg like a moonman on a spacewalk, until I reached the swim-up bar.

This vacation had been a reward to myself—for those days I arrived at the office at 6:00 A.M. and didn't leave until 9:00 P.M., for working on Christmas Day, for making myself care who won *The Bachelor*. For the first time in months, I'd started to relax. That was obviously shot to hell now. I needed to get out of my head for a while, and I needed reinforcements. Settling in on one of the submerged stools, I waved over the bartender who'd been taking care of us for the last few days.

"Okay, Hector, we have a *situation*," I said. "Bring the bottle of Jack Daniel's and a shot glass." I briefly relayed what had happened. He nodded understandingly and poured a shot for me and one for himself. We held our tiny glasses in the air.

Clink! The liquor burned a fiery trail down my throat. He immediately poured a second shot. Next I adopted a large family of piña coladas, forcing Hector to add rum until they turned brown. Forty minutes later Matt found me passed out on a lounge chair wearing Hector's baseball cap that said "Aruba: The bar is open!"

Three weeks later, I'd traded swim-up bars for coffee shops. Every day I went to some local café and trolled the classifieds for job openings. The economy had imploded seemingly overnight. Economists predicted the country was on the brink of a long recession—the Great Recession, they were calling it. No one was hiring. Not even the coffee shops. I'd already asked.

One morning I'd chosen a coffee shop where all of the baristas had facial piercings and tattoos. I got the impression they were judging me for ordering a latte. I placed my aging laptop on a table near the window and it groaned to life as though annoyed at being woken up at this hour. While the computer booted up, I snapped open the newspaper. A headline on the front page blared "80,000 Jobs Lost in March." I had been laid off in March.

It felt weird, doing nothing. I once spent fourteen hours a day cranking out blog posts and hysterically checking about thirty celebrity websites to stay abreast of breaking news. My BlackBerry had vibrated endlessly with gossip tidbits from fellow reporters. One time I took a ninety-minute flight and by the time we landed I'd received one hundred nineteen e-mails. When I wasn't at work, I was recovering from work. I felt so *available* most of the time that in my downtime I wanted to make myself as unavailable as possible. This meant going straight home after work every night, flopping onto my IKEA sofa, and watching people on television do the things that I was too tired to do myself. Within months, I was closely following the lives of about fifty fictional people, yet I had no idea what was going on with my friends. Even the *thought* of socializing had become exhausting. I'd started rejecting most of the invitations that came my way: brunches, birthdays, dinner parties, even a morning hike. Although I stand by that decision: friends don't make friends walk uphill before 11:00 A.M. I'd begun communicating primarily via e-mail, text messages, and Facebook status updates. I'd stopped wanting to meet new people at all. It was Matt who gently pointed out one night that I hadn't made a new friend in the three years we'd been dating.

"But I barely see the friends I already have," I'd sputtered. "I can't just go adding new ones to the mix or then I won't see *any* of them and I'll end up with fewer friends than I had in the first place!"

"Are you hearing yourself?" he'd asked.

"No," I'd replied, turning up the volume on the television.

For the past year and a half, Matt had been living in Albany, reporting on state government, so it had taken him a while to catch on to how much of a shut-in I'd become. I hadn't wanted him to worry about me, so sometimes when he called I'd turn up the TV about fifty decibels and shout into the phone, "Hey, babe! I'm out to dinner with friends! I'll call you when I get home!" I made up stories about what I was doing at night, and eventually I had trouble keeping my fake social life straight. What movie did I tell him I saw with my friend Jessica the other night? Whose birthday party had I supposedly gone to? I'd had to come clean after he caught me in a few lies and began to suspect I was seeing someone else. I'd told him I could never do something like that—it would require getting off the couch.

Matt thought that after losing my job, I'd use some of my endless free time to start socializing again. But your job is your currency in New York. "What do you do?" is often the first thing people ask upon meeting you. To tell people that you do nothing is like saying "I am nothing." It can actually stop conversations at parties. I'd rather skip those awkward exchanges altogether. Matt had been understanding, but I could tell he was weary of trying to haul me out of my apartment. He was tired of making excuses to his friends as to why I'd bailed out on yet another social occasion. I sensed he was waiting for me to return to the fun-loving, social person I was when we started dating. And that part of him worried this was simply who I was now.

These were the thoughts that occupied me as I stared at my computer. My screen, once so frenetic it could've induced epileptic seizures, had gone still. But that stillness was somehow more overwhelming. For the first time in my life, I had no idea what to do. Where did I go from here?

When I'd returned from Aruba a few weeks ago, I'd been ready to make a new life plan. I didn't want to blog about celebrities anymore. I'd enjoyed writing about A-list stars, but the celebrity landscape had changed in the last few years. More and more I'd found myself writing about reality stars, teenagers, and celebrities' babies. I was reminded of a conversation I had a few years ago. I'd been interviewing Joaquin Phoenix for a freelance article when he'd stopped me and asked, "Is this really what you want to be doing with your life? Writing about people who do interesting things instead of doing interesting things yourself?" Now, Joaquin went on to have something of a nervous breakdown. He grew a long beard, began wearing sunglasses indoors, changed his name to J.P., and quit acting for three years to pursue a career in hip-hop. Then he claimed the entire thing had been a "hoax." So he doesn't have a lot of room to criticize my life choices. Yet his question stuck with me. The truth was, I didn't mind writing about people who do interesting things. What I couldn't abide was spending my life writing about people who *don't* do interesting things.

So when I got back to New York, I'd created a Microsoft Word document titled "My One-Year Plan," where I could list my goals for the next year. No job meant my future was wide open. Too wide open, as it turned out. Weeks later, the document was still empty. Looking at the white screen that day, I felt I was looking at my future. Blank. The cursor blinked impatiently, like someone tapping a foot. I glanced again at that headline in the newspaper. I knew I was one of the lucky ones. No family to support. A degree from Yale. I'd gotten a pretty decent severance package and had some money in the bank to keep me going for a while. I had a wonderful boyfriend in possession of all his hair. I should have been rejoicing in the endless possibilities of my future. Instead I felt paralyzed, lost.

As soon as I logged on, an instant message popped up on my computer screen, breaking me out of my reverie. The merry IM tone echoed through the café, and I scrambled for the mute key. The message was from my friend Chris (a.k.a. GayzOfOurLives). As a blogger for *New York* magazine he was always online, so it had become a ritual for us to check in with each other every morning.

GAYZOFOURLIVES: *Whatcha doing?*
NOELLENOELLE: *Besides wondering who in my general vicinity has a wifi network called "penisface"? Nothing.*
GAYZOFOURLIVES: *Listen, I've been thinking about your state of affairs.*
NOELLENOELLE: *And?*

GAYZOFOURLIVES: *I believe you're having a third-life crisis.*
NOELLENOELLE: *A what?*
GAYZOFOURLIVES: *Well, you're too young to have a midlife crisis and you're too old to be having a quarterlife crisis. You're turning 29 soon. So, assuming you'll live into your late eighties, that would make this a one-third-life crisis.*

And there was that. My twenty-ninth birthday was the next week, and I knew my thirtieth would follow with startling alacrity. Yet another source of pressure. It can still be considered charming if you don't have your life together in your twenties, but when people find out you don't have some sort of direction by your thirties, they're a little embarrassed for you.

After signing off with Chris, I downed the dregs of my coffee and bellied up to the counter for a refill. While I waited my eyes drifted aimlessly around the café. Next to the register there was a flyer advertising free guitar lessons. Zany postcards were lacquered to the tabletop. Then something else grabbed my attention. On the far wall there was a small chalkboard sign featuring an inspirational quote of the day. Today's quote, written in pink chalk and squiggly handwriting, was:

> Do one thing every day that scares you.
>
> —Eleanor Roosevelt

"So what is it about this quote that's resonating with you?" my shrink, Dr. Bob, asked a few days later at our biweekly session. I'd started seeing Dr. Bob about a year before when I'd realized I knew more about Jennifer Aniston than I knew about myself.

"I don't know." I escaped his stare by looking around the room. It was full of nonthreatening furniture with rounded edges and warm taupe walls. The entire room was neutrals. Even Dr. Bob was neutral. His unjudgmental shrink eyes. He was neither tall nor short, not thin or overweight. His hair was somewhere between brown and gray and arranged in fluffy curls, making him look boyish, despite being in his fifties. He should have given up his practice for a life of crime. The man blended.

I was sitting—I always sat, never laid down—on his squishy leather couch, and he was across from me in his own chair. We were both slouching. It always felt like we should be watching the game instead of talking about my feelings.

"It's just . . . I used to *do things,* you know?"

"Do things?" he repeated.

"I don't try new things anymore. The older I get, the less I challenge myself."

He rubbed his fingers thoughtfully over his chin. "Can you think of a situation where fear stopped you from doing something?"

"Well, I don't know if I'd call this *fear,* but a few years ago I was at this karaoke bar with a bunch of my friends, right? I put my name in for a song—I think it was The Divinyls' 'I Touch Myself'—"

Dr. Bob raised an eyebrow.

"It was a lark!" I said defensively and continued. "Anyway, this guy went on right before me and did an incredible rendition of Journey's 'Don't Stop Believin'.' He had this whole dance routine to go along with it. At one point, he actually kicked over a chair! Everyone was going wild. Afterward people were still clapping him on the back, and then my song started up and I froze. I knew it was supposed to be all in good fun, but no way was I following that."

"So what did you do?"

"What do you think I did? I pretended it wasn't my song and I got the hell out of there."

"And what if you had gotten up there and bombed?" he mused. "What happened the last time you performed poorly at something?"

I mentally thumbed through the past few years of my life, but all I saw was work, a dinner here or there with Matt, the occasional summer blockbuster. "Okay, last year Matt and I went bowling and I threw ten straight gutter balls in a row."

"And what happened afterward?"

"And now I don't go bowling!" I said, exasperated. "It's not about *fear*. I just don't enjoy doing things I'm bad at."

"Avoidance *is* fear," he said gently. "When we're afraid of fear, we avoid situations that trigger it."

"But who cares if I don't bowl or sing karaoke?"

"The problem with avoidance is that it leaks over into other areas of our life. For instance, you've been avoiding new people, recreational activities, your friends—"

I interrupted, "That last one doesn't even make sense. I'm not afraid of my friends."

His voice remained patient. "No, but when our world feels out of control, we withdraw to maintain the illusion of safety."

I opened my mouth to object, then closed it. There was a moment of uncomfortable recognition.

He continued: "Fear can paralyze our lives. Fear of making the wrong decision keeps us from making any decision at all."

The empty one-year plan on my computer popped into my head, aggressive in its blankness. Was Dr. Bob right? Had fear slowly been consuming my life without my realizing it? My mind replayed all of the times I'd said nothing during work meetings because I was worried my idea would sound dumb. The time I'd turned down the opportunity to speak on a panel because I hate public speaking. How I'd stayed at bad jobs for too long because it was easier than leaving. Even small things like paying full price at flea markets because I was uncomfortable haggling with the vendors. How many chances had I squandered? How much of my life had been about avoiding life?

"Going back to this Eleanor Roosevelt quote," Dr. Bob was saying. "This could be a good project for you. You should run with this!"

"Huh?" I asked, snapping to attention. "Run with what?"

"Start doing more things that scare you!" When Dr. Bob got excited about something, his head actually bobbed. "You need to avoid avoiding. Practice confronting your fears," he said. "The more obstacles you overcome, the more empowered you feel, and the more you want to overcome other obstacles."

He had me nervous when he'd said *project,* but lost me completely at *obstacle,* a word that brought to mind climbing walls and tire mazes run by people in short shorts.

"Couldn't I just"—I cast about for an alternative and settled on an antianxiety medication—"take an Ativan or something?"

"Pills are a temporary fix," he said firmly. "What you need is a lifestyle change."

Lifestyle changes were for the morbidly obese. Or people who hoard newspapers until they're so walled in by the stacks that their living space is reduced to three square feet. A *lifestyle change*? I mean, get serious.

Dr. Bob's eyes scanned my face for a moment. "Noelle, anxiety can foster depression, impair your physical health, damage your relationships, and reduce your effectiveness in the world." His voice was concerned rather than matter-of-fact. "But if you allow yourself to fully experience fear, eventually you'll learn how to face it without being overwhelmed by it."

What could I say to that? I was trapped. If I refused, I was being an avoider. Dr. Bob didn't wait for my response. He knew it was better to let the idea percolate. Instead, he stood up, pulling the sides of his jacket together like curtains at the end of a play—the signal that our session was over.

"Think about this Eleanor Roosevelt idea," he said, opening his office door. "It might be the direction you've been looking for."

On my way home, I stopped by the Barnes & Noble in Union Square. It had an entire Roosevelt section, including a number of books written by Eleanor herself. I was skeptical about this whole fear-facing idea, but Eleanor had me intrigued. I was intensely curious about her life the way I once hungered for details about Angelina Jolie. I grabbed a couple of books off the shelf and plopped down on the scratchy industrial carpeting. Her life story was so rich that she wrote three autobiographies.

Skimming through her memoirs, I discovered that one of the most celebrated women in recent history was consumed by self-doubt as a kid. Her father, Elliott, doted on her, but one thing he had no patience for was her timidity. Because she adored him, she did her best to hide her fears from him. When she was six, the family took a trip to Italy. During a donkey ride through the mountains, Eleanor came to a steep downhill slope. She trembled with fright and refused to go forward. Elliott stared down at her and said, "You are not afraid, are you?" Fifty years later, Eleanor wrote, "I can remember still the tone of disapproval in my father's voice."

But it was her mother, Anna, who planted the first seeds of doubt. "I always had the feeling from a very young age that I was ugly," Eleanor said. She was forced to wear a back brace to correct a curvature of the spine, and she was painfully aware that her beautiful mother was embarrassed by her plainness. "I can remember standing in the door, very often with my finger in my mouth," Eleanor recalled, "and I can see the look in her eyes and hear the tone in her voice as she said: 'Come in, Granny.' If a visitor was there, she might turn and say, 'She is such a funny child, so old-fashioned that we always call her Granny.' I wanted to sink through the floor in shame." Anna suffered chronic migraine headaches, and Eleanor rubbed her mother's temples for hours.

"The feeling that I was useful," Eleanor later said, "was perhaps the greatest joy I experienced."

This was the problem with blogging, I thought. I'd been busy but I hadn't felt useful. One of the quotes in Eleanor's book that made me flinch was: "Great minds discuss ideas; average minds discuss events;

small minds discuss people." For years I'd been paid to write gossip about people.

I set down the autobiography and noticed the corner of another book peeping out from the pile. It was an unassuming little advice guide she wrote titled *You Learn by Living: Eleven Keys for a More Fulfilling Life*. The summary on the back read: "Offering her own philosophy on living, the woman who was called First Lady to the World leads readers on a path to confidence, education, maturity, and more." I flipped the book over and studied the photograph of Eleanor on the cover. She was in her forties, smiling gamely at the camera, swathed in a fur coat and a triple-strand pearl necklace. While not a beautiful woman, she was glamorous and confident, so different from the insecure child I'd just been reading about. I'd gone the opposite way, I realized. I'd been bold when I was younger but instead of challenging myself as I grew older, I'd simply eliminated the threatening things in my life. Deciding Eleanor's life story was something I needed to digest more fully, I gathered the autobiographies and advice book and headed to the cash register.

The next day at the coffee shop I blew through *You Learn by Living* in one sitting. Upon finishing, I turned back to the chapter titled "Fear— The Great Enemy" and reread it carefully. Eleanor credited fear as the great motivator of her life. "I was an exceptionally timid child, afraid of the dark, afraid of mice, afraid of practically everything. Painfully, step by step, I learned to stare down each of my fears, conquer it, attain the hard-earned courage to go on to the next. Only then was I really free," she wrote.

I leaned back in my chair, and my gaze fell upon the chalkboard. The Eleanor quote was no longer there, having been replaced with something by Maya Angelou. I'd memorized it anyway: *Do one thing every day that scares you*. If Eleanor determined her life purpose by conquering her fears, maybe it could help me figure out my future, with the added benefit of salvaging my friendships and reviving my relationship. Maybe to find out what I *did* want to do, I first had to do the things I *didn't* want to do. At the very least, conquering a fear each day would give me a goal to meet, a sense of purpose.

When I was reading *You Learn by Living*, I felt like Eleanor was talking directly to me: "The most unhappy people in the world are those who face the days without knowing what to do with their time. But if you have more projects than you have time for, you are not going to be an unhappy person. This is as much a question of having imagination and curiosity as it is of actually making plans."

"How long would you do it for?" Matt asked when I called him with the idea.

As a blogger I'd had deadlines every half hour, and my work had always felt careless and unfinished. Now I wanted enough time to fully devote myself to what I was doing. Time to ensure I wouldn't return to my old habits as soon as it was over, but not so much time that I'd burn out.

"I'll give it a year, starting on my twenty-ninth birthday." My thirtieth birthday seemed like a natural stopping place.

"A *year*?" he repeated, sounding incredulous. "I'm all for anything that gets you out of the house, honey, but have you thought this through? How will you support yourself?" The problem with Matt being a reporter was that he was always playing devil's advocate. The problem with me being stubborn was it made me dig my heels in even more.

"I can make enough money freelancing to stay afloat for a while— no one's hiring for full-time positions anyway—and I have savings." The more I defended the idea to him, the more invested I felt in it.

"Well, you know I'm here for you no matter what," he said, but his voice was dubious.

Chris was even less charitable. "It sounds a little crazy. I would hate to see you in a straitjacket, Noelle. They're really unflattering. They just add bulk." He added, "Though I have to admit, the idea of taking a year just to focus on me *does* sound pretty appealing."

Maybe it was crazy. Then again, our culture constantly sought life-style advice from celebrities, many of whom rose to fame on nothing more than sex tapes or a willingness to argue with others on camera while living in mansions provided by television networks. Wasn't *that* crazy? Eleanor was more than a celebrity—she was a role model. This was an anxious girl who grew up to become a social activist and a First Lady who held regular news conferences, wrote a newspaper column six days a week, and carried a pistol. In her downtime, she helped form the United Nations and establish the state of Israel. She assisted Franklin in carrying out the New Deal. It was an experiment in which the government poured resources into various programs to restore growth and public morale.

I told myself that this experiment could be my own New Deal: investing in myself now to create future growth. But part of me wondered if Chris was right and this was simply an exercise in self-indulgence. Shouldn't I be serving others? Then I remembered what Dr. Bob said about anxiety reducing our effectiveness in the world. Wasn't living a fearful life also self-indulgent? I wasn't fully contributing to the world if I was pulling back from it all the time. Not only that, worriers are draining to other people. I didn't want to keep dragging others down. If Dr. Bob was right about fear perpetuating fear in ourselves, my fears probably touched those I came in contact with in ways I couldn't even comprehend.

I went into Microsoft Word and opened up the blank document: "My One-Year Plan." Finally I knew where to begin. I started with the things Dr. Bob and I had talked about. The bowling and the karaoke . . . what was I really afraid of there? I wrote down, "Public humiliation. Failure." Then I thought about other things I'd been avoiding lately. My friends. Meeting new people. Public speaking.

"Rejection," I typed. Talking to my boyfriend about the future. "That Matt will leave me and I'll be alone." I paused and reread that line again, jarred by an anxiety I hadn't previously admitted to myself.

"Leaving this world with nothing to show for it but excessive knowledge of celebrity scandals." Thinking of my lifelong fear of an untimely death, I wrote, "Leaving this world before I'm ready."

Words were pouring out now, the cursor gliding easily across the screen as I attempted to list every fear I'd ever had, everything I'd backed down from or taken pains to avoid. When I finally stopped writing ten minutes later, I was astounded at the amount of sheer wussiness before me. The things I had listed ranged from physical fears (heights, flying, crashing into things) to more emotional fears (public speaking, criticism, confrontation, regret, disapproval) to the slightly ridiculous (sharks, sober dancing, the time I lied and told my dad I voted for McCain).

As I looked at the list, I saw how this could actually work. I really could confront a fear each day. Some of them could be grand gestures, like jumping off a cliff or skydiving; others could be small things, like telling someone what I really thought of them. Fear is relative. To some people, stepping on a stage is no big deal, but for me, the mere thought made my heart race. If something gave me butterflies or an inclination to flee, then it was worth trying. Just as I was about to pull up a calendar page on my computer and plot out my life for the next 365 days, I remembered something else that Eleanor said. I grabbed a book and flipped around until I found it: "You cannot use your time to the best advantage if you do not make some sort of plan," Eleanor wrote. But she cautioned, "I find that life is much more satisfactory when it forms some kind of pattern, though I do not believe in too rigid a pattern."

She's telling me not to go overboard, I thought. If you make this plan too rigid, then you're leaving no room for spontaneity, for facing the small, everyday things that come up unexpectedly. If I planned everything out, I wouldn't be facing all my fears, because in the last few years, I'd developed an aversion to spontaneity. I returned to my document and added one more line to the bottom of the list: "Fear of the unknown and unplanned."

Satisfied that I finally had, if not a plan exactly, at least a direction for my immediate future, I changed the document name to "My Year of Fear" and urged the cursor to the top of the screen. I clicked with more confidence than I'd had in months. A short message appeared across the screen: saved.

Chapter Two

↝

Nothing alive can stand still, it goes forward
or back. Life is interesting only as long as it is a
process of growth; or, to put it another way, we can
only grow as long as we are interested.

—ELEANOR ROOSEVELT

I felt hopeful for the first time in months. I also had a birthday to
plan. A few days after making the list, I was stretched out on the
couch reading an Eleanor book, but my mind kept drifting to my
upcoming twenty-ninth birthday. Because Matt would be at work in
Albany, he was taking me out to a nice dinner the next weekend. So it
would just be me and my few remaining close friends. This still left
the question of the celebration. Since my birthday was the first official
day of my Year of Fear, I wanted to combine the party with the scary
activity. But what?

Part of me had hoped Eleanor might inspire some ideas, but there
was no mention of birthdays as I thumbed through her biographies.
Instead I found myself sucked in to the drama of her privileged but
joyless childhood. Her parents' marriage was strained. Elliott drank
heavily. When Eleanor was five, he caused a bit of a scandal when he
fathered a child with one of the servants, who hired an attorney and

threatened a $10,000 lawsuit. When Eleanor was eight, her twenty-nine-year-old mother died of diphtheria. Elliott was in a mental institution trying to overcome alcoholism, so Eleanor and her two brothers moved into their surly grandmother's Manhattan brownstone. Five months later, her brother Elliott Jr. also died of diphtheria. Elliott and Eleanor mostly kept in touch via letters; one day the letters stopped coming. Less than two years after her mother's death, Eleanor's father jumped out of a window and killed himself. She and her little brother remained with Grandmother Hall and her four boisterous adult children who still lived at home. Her eccentric aunts—Maude and the unfortunately named Pussie—and her playboy uncles, Vallie and Eddie, were known for their wild shenanigans and love affairs. One day while the group was vacationing at their summer home, Vallie and Eddie parked themselves at an upstairs window with a gun and took turns firing at family members sitting on the lawn. Grandmother Hall pronounced the household too rowdy for a girl of fifteen and sent Eleanor to the Allenswood Academy, a finishing school for girls just outside London.

The headmistress at Allenswood was a silver-haired woman named Madame Souvestre who was not to be trifled with. She was French and required the students to speak French at all times. She demanded independent thinking from her pupils. Students who turned in papers that merely summarized her lessons found their work literally torn to shreds in front of the class, the pieces thrown to the floor.

"Why was your mind given to you, but to think things out for yourself!" Madame Souvestre cried.

I paused in my reading, trying to imagine my life at a school like that. At my high school, students had sometimes brought their children to class. One time someone dropped a backpack on the floor and the gun inside went off, hitting someone in the leg. School administrators couldn't figure out how to ban guns, so they banned backpacks instead.

Surprisingly, Eleanor flourished under Madame Souvestre. She engaged in lively discussions on world affairs. She tried out for field hockey despite having never seen a game and made the first team. "I think that day was one of the proudest moments of my life," she later said.

During school holidays, the headmistress invited her favorite student to accompany her on trips across Europe. Traveling with Madame Souvestre was "a revelation" for Eleanor. "She did all the things that in a vague way you had always felt you wanted to do." They took unmarked paths and changed their schedule on a whim. During an evening train ride through Italy, Souvestre spontaneously grabbed their luggage and ordered Eleanor off the train. She wanted to walk on the beach and see the Mediterranean in the moonlight.

"Never again would I be the rigid little person I had been before," Eleanor wrote of the experience. Her cousin Corinne barely recognized her when she entered Allenswood a few years after Eleanor. Her awkward, tentative cousin had blossomed into a confident young woman. "When I arrived she was 'everything' at school," Corinne said later. "She was beloved by everybody." Eleanor left Allenswood after three years at her grandmother's insistence to make her debut in New York society. That was the end of her formal education, though she vowed to never stop learning.

"Each time you learn something new you must readjust the whole framework of your knowledge. . . ." she said. "And yet, for a great many people, this is a continuing problem because they appear to have an innate fear of change, no matter what form it takes: changed personal relationships, changed social or financial conditions. The new or unknown becomes in their minds something hostile, almost malignant."

I have to learn something new. I put the autobiography aside. When I was little, I was always trying new things: new types of math, school plays, whatever sport the gym teacher decided to torture us with that day. While not always a success—dodgeball, I remember, being a particularly low point—I still tried. We didn't have a choice. Back then we had teachers and parents making sure we challenged ourselves. Then

I became an adult. The luxury of being an adult is you no longer have to do things that make you uncomfortable.

I dropped into my saggy armchair and brushed some birdseed off the armrest. The cage holding my parakeets, Jesus and Stuart, was next to the chair. When you live in a three-hundred-square-foot studio, everything is next to something. I leaned over my ottoman that doubled as a desk and fired up my computer. When my list of fears came onto the screen, I scanned it for ideas. Topping the list: *heights*. I did an Internet search of *heights* and *New York*. Forty-five million hits.

"Jesus!" I said out loud, then looked sideways at Jesus the parakeet. "Not you."

When I turned back, I noticed the Eleanor biography I'd left on my couch. Thinking of Allenswood Academy, I tried adding *school* to the search. The first website that came up was Trapeze School New York. Inwardly, I felt myself pull back, my go-to emotional reaction when faced with something unfamiliar. The company's slogan was, appropriately enough: "Forget the fear, worry about the addiction." I had to admit that it was perfect: a literal jumping-off point for my Year of Fear. I fought the urge to reach for a reason why I shouldn't do it. Instead, I picked up the phone and called Chris.

"So I've figured out what we're doing for my birthday." I told him all about the trapeze school. "I'm going to ask Jessica, too." Jessica is one of our closest friends, as well as the managing editor for Chris's website.

He gave a hollow laugh. "Can you call her when we're at work so I can watch? I need to see this."

"Very funny."

But Chris had a point. Jessica has the best body of anyone I know, but the last time she worked out was a few years ago when her local gym offered her a free trial. "That place may have actually been a prison gym," she reported afterward. "Touching the equipment pretty much necessitates a postworkout chemical bath. Never again."

She picked up before the second ring and I launched right in. "So for my birthday, I was thinking you, Chris, and I go to trapeze school."

"Oh Lord. Seriously? What are we—a *Sex and the City* rerun or something?"

"What do you mean?"

"There was an episode where Carrie takes a trapeze class."

"Okay, it'll be like that—but without the sex."

"Is there a bar at least?"

I paused to consider the question. Technically there was a bar, you just happened to swing from it.

"Yes."

Thus, on the evening of my twenty-ninth birthday, I arrived at Trapeze School New York flanked by two of my best—and most easily peer-pressured—friends. The school was an outdoor rig plunked on the roof of a five-story athletic complex on the shoreline of the Hudson River on the west side of Manhattan. It commanded a spectacular view of the city and the verdant fields below, full of people playing soccer and field hockey in the late afternoon sun.

"Ugh!" Jessica squinted at them and shook her head in disgust. "Look at all these healthy people! Forget the drug dens. *This* is the dark side of Manhattan." Jessica and I met four years ago through a blog. She was the editor of a site where I occasionally freelanced. Our friendship had been a slow build. It started over e-mail, moved to texting, and then progressed to in-person appearances. But now not a day went by that we didn't talk.

She gathered her freshly highlighted brown hair into a ponytail and pulled it tight with grim resignation. The night before she'd called to report that the closest thing she had to workout pants were last season's leggings. They were meant to be worn under dresses and had a see-through crotch. "Do you think I can wear them anyway?" she'd asked. "I'm confident in the appearance of my vagina." Luckily, someone at the office that morning happened to have an extra pair of yoga pants.

"Nice shirt, by the way," I told Chris. He was wearing a body-hugging women's T-shirt reading: BEAT ME WITH 10 POUNDS OF VOGUE.

He grinned. "Hey, the website said to wear fitted clothing." He looked gawky, but Chris was more athletic than Jessica and me combined and squared to the power of infinity. Although I wasn't the worst player on my high school soccer team, I was the only one to accidentally score a goal for the opposing team. Chris, meanwhile, had grown up in Maine doing Maine things like hiking and snowshoeing. He spent a summer biking across the United States for Habitat for Humanity. He was on the crew team at his prep school and at Yale. He hadn't aged a day since we'd met ten years before working for the *Yale Daily News*: same meticulously clipped blond hair and high cheekbones set in a warm, appealing face.

As we latched on our safety belts, I gazed up warily at the trapeze rig. It looked like your average circus trapeze, except that the aluminum ladder leading up to the platform was startlingly rickety. Stretching beneath the entire monstrosity was a large net, which somehow provided no comfort. I pictured a cartoon version of myself falling through the net and barreling past the center of the earth to China, where I'd pop up in a bowl of chow mein to the surprise of a chopstick-wielding diner. For Matt's sake, I was glad he wasn't here. His fear of heights far outstripped mine. On our first Valentine's Day he'd taken me to the observation deck on the seventieth floor of Rockefeller Center. He'd stood behind me the whole time with his arms wrapped around me and months later confessed he'd been clutching me out of terror, not affection.

When I think of the flying trapeze, I think of the circus, so I'd arrived expecting a certain level of merriness. But trapeze school was conducted in a businesslike manner, the instructors bordering on brusque. Our ground instructor was a thirtysomething Adonis named Ted with abs like a cobblestone street. He taught us the proper way to handle the trapeze. When he yelled "ready," we were to bend our knees; when he yelled "hep" (trapeze speak for "go!"), we were to take a tiny hop forward off the platform.

"The trapeze bar is always heavier than you'd think—like an Academy Award—so be sure to lean back when you grab it or it will pull you forward," he said.

"Give him an Oscar for Best Achievement in Abdominals," Jessica whispered.

Ted continued: "There won't be any practice runs where you just swing out and say 'Wheeeeeee!' You're going to do a trick called a knee hang. On the first swing out we want you to pull your knees through your hands and hook them over the bar the way you did when you were kids on the playground. Then you will let go with your hands and—hanging by your knees upside down—stretch your arms out in front of you and make the Superman pose. On our command, you'll grab the bar again, unhook your knees, and hang straight. Then you'll do a backflip dismount and land on the net on your back."

"Heh!" I let forth a short, disbelieving laugh and murmured to Jessica, "For the record, should something go wrong and I end up on a ventilator, pull the plug."

"Keep me on life support," she said. "I spend most of the day in a vegetative state anyway. Does it really matter if it's at the office or in a hospital bed?" She paused. "But if there's permanent damage to my face, do *not* resuscitate."

Trapeze was a two-instructor operation. Ted and his award-winning abs worked the safety lines; a man named Hank ran the platform. There were seven other students in our class. Our order was determined by when we signed in. Happily, I'd signed in last. First up was a sixteen-year-old gymnast. She executed the knee hang flawlessly with pointed toes and fabulous extension. Everyone clapped except Jessica, who muttered under her breath, "Bitch, get your ass to the intermediate class. Like I don't feel bad enough about myself as it is?"

But when it was her turn, Jess easily performed her knee hang and backflip. I was surprised and not surprised. Jessica is a person of incongruities. Petite with a big personality. Sweet-looking face and a bit-ing sense of humor. She's the most opinionated person I know but my least judgmental friend. A New York sense of style and a down-home Michigan accent. Chris's flip was less elegant, since he had longer limbs to contend with, but he did well on the trapeze too.

"Nicely done," Jessica said when he retook his seat next to us.

Gesturing to the swings and harness, he replied, "It turns out, I've had enough gay sex to prepare me for this experience."

When it was my turn, I dipped my hands into the basket of powdered chalk at the base of the ladder, per Ted's instructions. The chalk soaked up the sweat and kept your palms from slipping.

"Ten pills of Xanax says she doesn't jump," Jessica stage-whispered to Chris.

"You're on," Chris replied.

"I heard that!" I called out.

I crept up the ladder as slowly as possible while still being in motion. Seeing my white, disembodied hands clutching the rungs, I remembered something a surgeon had said in a documentary I once saw on TV. When reattaching a hand, the hardest part is not setting the bones or connecting the arteries—it's fixing the nerves. Once severed, nerves are not easily restored. After surgery they grow back at a rate of one inch per month. It's a long process. Sometimes they never come back and the hand will be forever paralyzed.

Jessica had told me that the scariest part was going up the rickety ladder, but once I climbed onto the platform I knew that she was a damned liar. Since the rig was atop a five-story building, the trapeze seemed to reach unearthly heights. I stood up on quaking legs and immediately latched on to a nearby metal pipe that looked relatively secure. Waiting for me on the platform was Hank, the sixtysomething instructor who ruled his aerial fiefdom with a firm hand and a mustache normally reserved for sheriffs in 1940s westerns. After a curt "hello," he briskly clipped the safety lines to my waist harness. I hoped he didn't notice that the back of my tank top was soaked through with sweat.

"Now then! Here you go!" he boomed, holding the trapeze in front of me. I didn't reach for it. I could tell by his no-nonsense expression that he was going to try to shame me into grabbing it. "C'mon now, it's just like stepping off a curb. You're not afraid of stepping off a curb, are you?"

I didn't know what curbs were like in his neighborhood, but mine didn't include thirty-foot drops and signing a waiver beforehand "in the event of death or accidental dismemberment."

I tugged suspiciously on one of the ropes attached to my safety harness. "Is there any danger of getting tangled up in these on the way down? Could I decapitate myself or something?"

"It hasn't happened yet," he said, and I swear I detected a note of hope in his voice.

I looked over my shoulder toward the ladder and sighed. The only thing scarier than jumping off this platform was the prospect of going down that ladder backward. I wondered if they'd made it rickety on purpose to prevent people like me from backing out. I decided that I'd be okay with remaining on this coffee-table-sized platform for the rest of my life. I'd make it work. I could get a job manning the platform like Hank. I could get every meal delivered. "I live on the top floor," I'd tell the delivery guy.

"Just think of it as stepping off a curb," Hank repeated, less patiently this time.

"A *curb*?" I snapped. "Where do you live—with the Jetsons?" Bons mots were my preferred defense mechanism, though I would use homicide in a pinch.

Hank's ongoing patter was becoming part of the background noise like the traffic. But there was no ignoring the basic truth that the longer I hesitated, the more loaded the moment became. For the rest of class, the students and instructors would view me in the context of this moment. Yet, I couldn't seem to make myself move. Minutes were passing, which I could tell only because my eyes flickered to the ground every so often and each time, a different emotion was visible on the faces of my classmates staring up at me. Encouragement was replaced by pity, impatience, then irritation. Only Chris's and Jessica's expressions remained unchanged. They looked just as hopeful as they had fifteen minutes before. And in the end, this was what compelled me to reach out for the trapeze. To watch them lose faith in me would have been awful. *You don't necessarily have to jump,* I told myself. *You just have to lean far enough forward that you can't go back.* Gravity would do the rest.

A relieved cheer went up from the class as my hand grabbed the metal. Having long forgotten the Academy Award simile, I was caught unawares by the heaviness of the trapeze. The fifteen-pound bar yanked me forward, and out of instinct I dropped it. It sailed forward and with nothing to grab on to I teetered on my tiptoes, windmilling my arms. Gravity did the rest, and I careened off the platform. Ted, who was on the ground controlling the safety lines, jerked on the cables so I fell only about four feet.

"Quit moving your arms and legs!" he commanded. I went still and floated in midair, a marionette waiting for instructions. Ted reeled in the lines, and I rose up a foot at a time. When I was level with Hank again, he hauled me back onto the platform by the back of my waistband harness.

He *tsk*ed and shook his head. "Are you ready to get serious now?"

"I'm seriously ready to get *down* now."

He dragged the trapeze back to me using a giant hook that looked, appropriately, like the kind used to yank vaudevillian performers off the stage when they'd overstayed their welcome.

"Just listen for the commands," Hank reminded me. "When Ted says 'ready,' that's your cue to bend your knees. When he says 'hep,' jump off the platform. Got it?"

I gathered myself and nodded firmly. "Got it."

"Ready!" Ted yelled.

I bent my knees.

"HEP!"

I didn't move.

"Do we need to go over this again?" Hank asked.

"I'm sorry, I'm sorry. I just got spooked. I'm ready now."

Hank nodded at Ted again.

"Ready! . . . HEP!"

I took a bunny hop forward. I couldn't describe what it was like to take that first swing out, mostly because my eyes were squeezed shut.

"Open your eyes, open your eyes!" Ted shouted from the ground.

When I forced open my eyelids, I saw that I was traveling faster than I'd anticipated. A lot faster. It was exhilarating and dreadful. As I hurtled forward, Ted screamed at me to hook my legs over the bar.

"Your knees, your knees!" he shrieked.

My ass, I thought, but surprisingly the backswing gave me enough momentum that I hooked them with ease.

"Now let go with your hands!" Ted yelled when I reached the height of my arc.

This was the part I'd been worried about since the beginning of class. I'd been afraid that once I let go, I wouldn't have the strength to pull myself up when it came time to grab the bar again. I'd just have to hang there. I'd be like those bears that break out of the forest, climb utility poles in suburban neighborhoods, and cling for dear life until they're shot down with a tranquilizer dart.

Gritting my teeth, I released my hands. The act of letting go—unfurling my body and falling back into nothingness—was slightly liberating, but also extremely unnerving.

"Arch your back, arms out!"

I held my arms out in front of me, Superman style. On the return swing, I glimpsed an upside-down Hank giving me the thumbs-up from the platform. Or maybe it was the thumbs-down? Before I could figure it out, it was time to grab the trapeze again. I curled toward the bar, and when my hands found the metal, I clamped my fingers down as tight as I could. Then I unfolded my legs so that I was hanging straight again.

"C'mon, Noelle, it's time for the dismount!" Ted called. As the trapeze charged forward a third and final time, I pulled my knees up to my chest, let go, and did a perfect backward flip into the net below. I also whacked my toes on the bar so hard that the scream is still echoing in the Catskills. But other than that, I did pretty well.

A few days before, during my session with Dr. Bob, I asked him, "Why am I afraid of heights?"

"Because you're smart!" He laughed. "Take a look at people's top fears: snakes, insects, rats, and heights. Evolution has programmed certain fears into our brains to keep us alive. Fifty thousand years ago, people steered clear of snakes, insects, and rats because they carried diseases. Our distant ancestors who were afraid of heights didn't fall off cliffs."

"But I thought fears were learned?"

He shook his head. "Some fears are learned, some we're born with. Get this: There was a study where psychologists placed an infant on a table with a pane of Plexiglas in the center. Now, the baby could easily crawl across this Plexiglas—but almost all the kids refused. Why?"

"Because the Plexiglas made it look like they were going to fall."

"Kittens and puppies also refused to cross the glass," he said. "Then they brought in some baby ducks. Guess what? The ducks walked across without a quack of protest. Now why weren't the baby ducks afraid?"

"Because they have wings?" I venture.

"Exactly."

I think about this for a moment. "But if a fear is instinctual, aren't we just . . . stuck with it?"

"If we can experience a seemingly risky situation over and over without harmful consequences, we can train our brains to be less afraid."

"So you grow your own wings, basically."

"Exactly."

After I limped my way back from the trapeze rig, Chris kindly lent me his cold water bottle, which was lying horizontally across my swelling toes. My cheeks were flushed with excitement over what I'd accomplished, but I was also more scared than I'd been at the beginning of class. Before my turn on the trapeze, there had still been a chance it wouldn't be scary, but now I had confirmation. Now I knew exactly how fast and high it was going to feel. But I also knew that while it was brave to do something you think is going to be scary, it was braver to do something you *know* is going to be scary. And to have faith that, eventually, it will start to be less scary.

This actually turned out to be true. After I had performed the knee hang and backflip combo three more times, my heart rate had slowed considerably. After the fourth turn, my hands stopped shaking. Right before our final turn, I noted with trepidation that a stocky Latino gentleman was sitting on the second trapeze at the other end of the rig. At which point Ted announced that it was time for us to do "the catch."

"You'll swing out and hang by your knees just like before. But this time, when you put your arms out, my man Pepe here"—the man on the second trapeze, now hanging by his knees, gave a friendly upside-down wave—"will catch your hands. Then you'll let go with your legs and he'll swing you out across the net."

A nervous tingling simmered in my stomach. I didn't understand the mechanics of this operation. What if he grabbed my hands but I didn't release my knees in time? I pictured myself ripping in half—my arms and torso carried off by Pepe while my legs and knees, still hooked over the bar, swung back toward Hank.

"I don't know if I can do this," I whispered. Jessica looked equally uncertain.

"I don't know if I can hold hands with someone named Pepe," Chris said.

The gymnast was the first to go. Because she was more advanced, Ted had given her a more complicated stunt to perform. Instead of doing a knee hang, she did a perfect upside-down split. I held my breath as she took her hands off the bar and reached confidently for Pepe.

"Oooh, aren't we special?" Jessica grumbled as the girl backflipped onto the net. "Sure, she can do tricks, but can she menstruate?"

As I climbed the frail ladder a final time, my terror had downshifted to mere apprehension. It helped that the sun had gone and the rig was now illuminated by spotlights. The effect was festive, like a real circus, but more important, my world had been made smaller, less overwhelming. Instead of looking around, fretting over how high I was, I could only see what was right in front of me. I focused on each individual ladder rung, the meditative rhythm of my hands, and made it quickly to the top this time. When I pulled myself up onto the platform, Hank looked impressed.

"We'll make a flyer out of you yet." He grinned and I grinned back. I took the trapeze from his hand. On the other trapeze, Pepe was hanging by his knees, building up momentum. I tried not to imagine accidentally crashing into him. Instead, I assumed the position, leaning back, toes hanging over the edge of the platform.

"Ready!"

I bent my knees in anticipation. The tension was almost unbearable in its very *in-betweenness*. It was the pause at the top of the roller coaster when it's no longer going up but not yet going down. It was the moment after the track runners have taken their marks, but just before the horn blares. It was a moment between moments, defined by what's happened before it and what is about to happen. It was nothing, but it was everything.

"HEP!"

I sliced through the air, enjoying the sensation of it whooshing past my ears.

"Hook your knees, Noelle!" Ted's voice called from below. Using the last of my abdominal strength, I drew my legs to my chest and hooked them over the bar. Back arching! Arms extending! Here was Pepe! His meaty hands clamped firmly onto mine. I straightened my legs and my knees released the bar quite naturally. Now I was no longer upside down but soaring toward the twinkling New York skyline. I'd never been one of those people who described cityscapes as *beautiful,* but it was dazzling. Millions of tiny windows glowed crisply in the darkness. The class cheered—no one louder than Chris and Jessica—and someone let out a whistle. I flopped down onto the net and staggered toward the edge, grinning goofily. I felt the stirrings of something I hadn't felt in a while: pride. Not the pride that comes from a salary bump or a promotion, but the kind that comes when you've pleasantly surprised yourself. At the end of class we said our good-byes (Hank actually gave me a high five) and gathered our belongings.

"Now can we go get drinks?" Jessica asked.

"Absolutely. I'm buying," I said.

As the three of us limped through the gate on sore legs, Chris asked, "So you're really going to do three hundred sixty-five different scary things this year?"

I shook my head. "No, the quote only said that I have to do one scary thing every day. It doesn't have to be a new thing each time. If we learned anything today, it's that just because you've done something frightening once doesn't mean it's suddenly not scary anymore."

"Still," Jessica said doubtfully, "scaring the crap out of yourself every day? I don't know if I could make it a whole year."

When she said this, a new thought started nagging at me. I'd been so focused on making it through this one day that I hadn't considered how it would feel to do this day after day after day after day. Looking at my cell phone, I realized it was already 10:00 P.M. Just nine short hours before I'd have to wake up and face my fears all over again. I swallowed but said nothing. I wasn't sure I'd make it either.

Chapter Three

Looking back I see that I was always afraid of something: of the dark, of displeasing people, of failure. Anything I accomplished had to be done across a barrier of fear.

—ELEANOR ROOSEVELT

"So it seems my Ivy League–educated daughter has run away and joined the circus?" The voice on the phone was joking but concerned.

I groaned inwardly. "Hi, Dad." I knew I shouldn't have e-mailed him the pictures from my trapeze class. This was a man whose favorite coffee mug read: *Sometimes the road less traveled is less traveled for a reason.* But my parents had been so skeptical about the entire Year of Fear project. I thought the absurdity of the trapeze class might warm them to the idea.

There was a fluttering in the background and I knew my dad was flipping through the *Wall Street Journal,* which he's read front to back every day for the last twenty-five years. (When he found out what paper Matt wrote for, my dad sniffed, "Well, I hope he's not one of those elitist liberal kooks. That paper's full of them. See, what I like about the *Journal* is that they're not biased.")

"Did I catch you at a bad time?" he drawled. "Busy trainin' for the ice capades?"

"I'll have you know, I'm at a coffee shop, working on a freelance article for a magazine."

"Have you thought any more about law school?" he asked. Leave it to Dad to dive right in. "Now *that's* job security right there. You could make $300 an hour writing wills."

"And then I would die of boredom, which would have an agreeable symmetry." With my free hand, I started massaging my temples to ward off the headache that always followed this debate. My dad was a businessman, specializing in fiber optics, whatever that means. When I was a kid, he was always dragging me to the office to groom me for corporate life. I'd usually pass the hours Xeroxing my face. Once he'd realized I wasn't destined for business, he latched onto the idea that I should be a lawyer and hadn't let go for two decades.

"Well, I think you should consider moving out of that city and coming home to Texas." My parents viewed my living in New York as if I was studying abroad or on some kind of caper. I sensed they were awaiting my return to "real life."

"New York is my home," I said firmly.

"Well, it's an awfully expensive place to live when you don't have a full-time job. How much are you payin' for that apartment these days?"

"Bye, Dad!"

I laid my head down on my keyboard and banged it a few times, causing *jfkdjfkdjflkdjfdlkjfdlksjfdlsjf* to jump onto the screen. When I went to delete it, I saw that I had an e-mail from my friend Bill. It contained only one line: *Want to rage in the cage with me this weekend?*

"Excuse me?" I wrote back. "Are you challenging me to a steel cage match?"

"Close," he replied. "I'm going shark cage diving this weekend. Sharks are on your list of fears to conquer, right?"

I hesitated. Sharks are a long-held fear of mine stemming from a

1986 home screening of *Jaws*. The ocean, I'd suddenly learned, was full
of beasts that killed indiscriminately and with musical accompani-
ment. Warily, I clicked on the link Bill sent me. It was a website for
a shark cage diving company called Happy Manatee Charters. It was
hosting a two-day expedition this very weekend. The boat left on Friday morning and returned late Saturday afternoon.

Since taking that first step off the trapeze board three weeks before, I'd taken on one fear a day, per my mission, but they'd been small
victories, things that I would've let slide before, but not under the new
Eleanor Roosevelt administration. I'd sent my salmon back at a sushi
restaurant for being too fishy. I'd called up my credit card company
and asked it to lower my interest rate, and after speaking to four different supervisors, they finally agreed. Matt and I had gone to a sold-
out movie and were pleased to find, in the packed theater, one row that
was completely empty except for a single college-aged guy sitting in
the center. Apparently, he'd been sent ahead as a representative for
less punctual friends because when we went to sit down, he called out
smugly, "This entire row is saved, bro." We turned to keep walking up
the steps; then I stopped.

"Not anymore!" I declared. Over the guy's sputtering protests, I
plopped down in one of the forbidden seats, pulling Matt down into
the seat next to me. Normally I would've slunk away and stewed about
it for a while. It was nerve-racking but exhilarating.

I'd realized in the previous weeks that I needed to be practical
about this experiment. If I was going to keep this up for a year, not
every challenge could be as elaborate or expensive as trapeze lessons.
Some fears had to be faced in small moments, whether preplanned or
things that just came up during the day. When I started paying attention, I saw how often I avoided confrontation. I'd told myself this was
a sign of maturity. After all, wasn't it childish to make a fuss about
trivial matters? Now I realized that at the heart of these situations was
a fear of offending someone. But if I couldn't stand up to people when
there was little at stake, how would I summon the courage when it did

matter? Standing your ground could be even scarier than standing on a two-story platform.

I didn't feel ready to face sharks yet. There was only a month left of summer and I'd been hoping to put sharks off until next spring so I could work my way up to it. But maybe Bill's e-mail was a sign that it was time for another big challenge. Besides, putting it off would be another instance of me avoiding confrontation. And to quote Eleanor, "What you don't do can be a destructive force."

Screw it. "I'm in," I e-mailed back.

On Thursday afternoon, I boarded a three-hour train to Greenport, Long Island, where the boat, *The Manatee,* was scheduled to depart the following morning. Upon arrival, I checked into a cheap motel room and called Bill.

"There's a beer bottle opener nailed to the bathroom wall in my motel room. Jealous?"

"That room obviously knows what you're going to be facing tomorrow," he said. "Sorry I can't be there to help you put it to good use."

Bill cohosted a nightly television talk show and couldn't take off work Friday. So as soon as he was off the air, he'd jump in a rental car and meet our boat at Martha's Vineyard late Friday night.

"It'll be nice to see you, Hancock," he said. "What's it been—seven, eight months?" Wow, had it been that long? Like so many of my friendships lately, ours had been subsisting on e-mail and texts, a sort of friendship life support system.

Bill, in particular, had been totally behind the concept for a Year of Fear. He and I had met ten years before during my summer internship at *Stuff* magazine. He was the features editor, and as soon as I saw that he kept an inflatable alligator raft on the floor next to his desk, I thought, *I need to be friends with this man.*

"So are you prepared to die a horrible death tomorrow, Hancock?" he teased.

"Don't remind me." I groaned. "I can't believe you actually do this stuff for fun."

Actually I could believe it. He was a direct descendant of William Dawes Jr., who accompanied Paul Revere on his famous midnight ride. Fearlessness was practically in his DNA. This was a man who, in his late twenties, sneaked into a college and posed as a student for a week to see if it was as fun as he remembered. He finished the Walt Disney World marathon in Orlando. This wasn't particularly unusual except that he did it in drag, changing every five miles into a different Disney princess costume.

When we signed off, Bill said, "I'll see you on Saturday." Then, pausing dramatically, he added, "Hopefully." I hung up on his cackling, movie villain laughter.

This adventure seemed especially apropos considering water was one of Eleanor's longest-running fears. It started when she was three years old, traveling by steamship to Europe with her parents. On the first day of the sail another boat got lost in the fog and collided with their ship, killing and injuring passengers. Elliott and Anna Roosevelt fled to a lifeboat while Eleanor stayed on deck. The plan was for the crew to drop her overboard and Elliott would catch her. "My father stood in a boat below me, and I was dangling over the side to be dropped into his arms," Eleanor wrote. "I was terrified and shrieking and clung to those who were to drop me." She screamed as she fell through the air and into Elliott's waiting arms. Water and heights, unsurprisingly, fell out of Eleanor's favor after this encounter. It didn't help that Anna and Elliott sent their traumatized daughter to stay with relatives while they continued on their six-month tour of Europe. From then on they usually left her behind when they traveled because of her fear of boats. Being fearful, Eleanor learned, came with consequences.

A few years later there was another unfortunate incident while visiting cousins in Oyster Bay. Uncle Teddy Roosevelt "was horrified that I didn't know how to swim so he thought he'd teach me as he taught all his own children, and threw me in," Eleanor recalled. "And I sank rapidly to the bottom. He fished me out and lectured me on being frightened."

The next morning I walked to the dock and met *The Manatee*'s captain, Gus, a hefty man with long brown dreadlocks and a monotone voice. The fishing boat was smaller and more rustic than I'd expected. The sleeping quarters were belowdecks in the hull. There was only room for two people to stand up at a time. There were bunk beds—bunk benches, really—built into the slanted walls. There was no shower on the boat, just a hose on deck with water pressure that would be the envy of any fire department. And while I've never considered myself someone who demands the finer things in life, I do enjoy a good roll of toilet paper, and there was none in the bathroom—or "the commode," as Captain Gus called it. I dug my cell phone out of my backpack and called Bill from the deck to ask him to bring toilet paper. Then, after making sure no one was within earshot, I hissed into the phone, "What if I have to go number two before we dock at Martha's Vineyard?!"

"Maybe you're supposed to use the hose on the back of the boat," Bill answered cheerfully. "Just think of it as an industrial-strength bidet."

All three of the other people on the trip were experienced divers. I took an instant liking to Ronald, a retired lawyer whose T-shirt read: WORK SUCKS, I'M GOING DIVING! On the back he had written BITE ME! in black marker and drawn a shark face, portrayed in three-quarter profile. Les, an underwater photographer, was a good-looking blond guy, but there was something creepy about his manner that I couldn't quite explain. Mandy, a special-ed teacher from Pennsylvania, took off her swimsuit cover-up to reveal a neon pink bikini and an assortment of tattoos. Stretching across her lower back was an underwater seascape featuring sea horses, coral, and sea turtles. There was also a scuba diver tattoo on her left shoulder and, for the sake of variety, a mouse riding a motorcycle on her right.

"They were done by my friend Mona. She really is an artist!" she said, arranging her body on the deck for optimal sunbathing.

The plan was to make the six-hour journey over to Martha's Vineyard — stopping along the way to do some shark cage diving—and dock for the night. Then we'd repeat the journey, in reverse, the following day. The motor coughed to life, and soon Gus was navigating us out of the hamlet, steering with his right foot while eating a bowl of Cheerios. My stomach flip-flopped. There's nothing more nauseating than the sight of milk in ninety-degree weather. An hour into the sail, I was downright queasy. I'd never been seasick before, but I knew my day had come as I watched Gus toss pieces of cut-up fish (called chum) into the boat's wake to create a slick for the sharks to follow. Once the slick was several miles long, we dropped anchor and Gus pulled out a sealed, perforated bucket full of frozen fish.

"What's that for?" I asked in a weak voice.

Gus tied a rope around the handle and pulled on it a few times to make sure it was secure. "I'll throw it overboard and the smell will leak out of those holes and attract sharks." He tapped the bucket fondly. "I call it my chumsicle."

"Sharks can smell blood up to a mile away and—" Ronald added, but was cut off by my heaving my bacon, egg, and cheese breakfast sandwich over the railing.

"Great!" Les said with a leer. "More chum!"

Ronald patted my back sympathetically.

Most of the deck was taken up by the four-by-four-by-seven-foot shark cage that Gus had welded himself. The cage could hold two people at a time, and the top operated on a hinge so divers could get in and out. The bars were six inches apart. When Gus first built it, he'd left a three-by-three-foot hole in the middle so people could photograph the sharks without having bars in the way. Then one day a blue shark swam into the cage with one of the divers and started thrashing around. No one was hurt, but afterward Gus filled in the space with bars that are about eight inches apart.

"So have any sharks gotten in since then?" I asked.

"Just a few mako sharks." Gus shrugged. "But they usually just swim right out."

Usually?

"So, uh, what do these makos look like?" I asked, but Gus was busy lowering the cage into the water at the back of the boat.

"They have long slender bodies," Ronald piped up. "That's how they can fit between the bars."

Les held out his camera. "Here's a picture of one." It was fairly short, by shark standards. Its long thin snout gave the impression it was accusing you of something.

"That looks like my old boss," I said.

SPLASH!

All eyes turned to the water where a dark silhouette glided ominously just beneath the surface.

"A blue shark," Ronald observed. "About ten feet long, I'd say."

"Noelle, come over and shark wrangle while I get the cage ready," Gus said.

Wrangling entailed dangling a piece of string knotted with fish into the water, then when the shark went for the bait, pulling the string out of the water the way a matador rips the red flag from in front of a bull. It was pretty fun, actually. This move was designed to keep the shark near the boat, instead of taking the bait and flitting off. But the blue kept darting away.

"Am I doing it wrong?" I asked.

"Something is spooking her," Ronald said knowingly. "Usually that means there's a mako nearby."

"Why would a smaller shark make the big shark go away?"

"Mako sharks are the fastest fish in the ocean and very aggressive. Sometimes they eat other sharks. In fact, mako embryos sometimes consume each other for nutrients while gestating in their mother's body."

"They eat their brothers and sisters before they're even *born*?" I asked. "That's cold."

This biological detail didn't faze Ronald, who was already climbing into the cage with Les, eager for some face time with Big Blue. She never returned, however, and after a half hour in the water, the two of them came back up looking disappointed.

"Dude, I told you guys when you signed up, there's no guarantee how many sharks we'll see on these dives," Gus said, his tone defensive. "Sometimes there aren't any at all. I just do what I can and hope for the best."

Les started telling me personal stories to take my mind off the seasickness. He recounted the time he'd slugged his daughter in the face "to show her who was in charge," at which point I announced, "I think I'll go down in the cage with Mandy." I snapped myself into a wetsuit and tried not to think about my lack of diving experience. Technically you don't need to be certified if you aren't going below twelve feet of water, but I'd never breathed with a respirator before. My mind kept returning to a *Choose Your Own Adventure* book I'd read as a kid. The main character in the story was a scuba diver on the hunt for sunken treasure, and at the end of the story the reader got to choose his fate. You could choose to risk the small amount of oxygen left in the tank and go for the treasure. Or you could play it safe and return to the boat, but risk never finding the treasure again. I kept my finger on the page and skipped ahead to see what happened. The one who took the chance ended up running out of air and suffocating on the sea floor. The one who turned back went on to live a happy, but presumably boring, life.

Gus gave me a three-minute tutorial on the correct way to use an air hose and how to empty my mask if water seeped in. My stomach turned over. This time it was nerves and not seasickness. Wait—the seasickness!

"What if I throw up underwater and start choking?" I asked.

"Vomit into your respirator," Gus advised, and I looked at my respirator in disgust.

"What do I do if I'm in trouble and I need to come up?"

"Signal me by opening the top of the cage. I'll see you from the surface and reel you in," he said. "But *don't forget* to close it immediately."

"Why?"

"If the door is flapping open, I can't grab the cage and lock you into the back of the boat when you get to the surface."

"What happens then?"

"Then the cage goes underneath the boat," he said.

I shuddered.

He added: "Oh, and keep an eye on your hands when you're holding on to the bars. That's an easy way to get bit."

Once Mandy and I were securely in the cage, Gus shut the top, tied it closed with a bungee cord, and lowered us down twelve feet. Soon we were riding on an underwater roller coaster without seat belts. The choppy water jerked us back and forth, and we had to hold on to the bars to keep from crashing into the walls and ceiling. Mandy and I stood back-to-back. That way, if a shark approached, one of us would see it coming and signal the other. I clenched my respirator so hard my lips ached and then lost feeling entirely. Breathing underwater felt claustrophobic, as if the water actively wanted to get inside my body.

Mandy grew bored after twenty minutes of no action and signaled to Gus. He pulled the cage up to the boat to let her out.

"Had enough?" he asked me.

"I'll hang out for a while." The old Noelle would've quit while she was alive and followed Mandy, but I was determined to see this through.

"Rock on!" He flashed me the "sign of the horns" hand gesture and lowered me back down.

Now I stood in the middle of the cage, arms gripping bars on opposite walls to keep steady. I canvassed the murky water in front of me.

I looked over one shoulder, then the other. I checked on my hands to make sure they were still there. Visibility was low. I wouldn't be able to see the sharks until they were pretty close.

Suddenly, a flash of tail in the distance. Then nothing.

Oh shit.

I once read that Steven Spielberg had technical difficulties with the mechanical shark while making *Jaws*. The animatronic fish (nicknamed "Bruce") kept shorting out, delaying production, so Spielberg compensated for Bruce's absence by using it to create tension. Now I understood why his technique worked: not knowing was worse than knowing. The scariest parts of the movie were the times when you didn't know where the shark was but felt its presence, loitering in the shadows.

Then I saw it, about seven feet long and winding its way over to the fish Gus was dangling in front of my cage. A mako. Naturally, this was the shark that showed up on my watch. When it was about a foot away from me, Gus yanked the fish out of the water. The mako, frustrated at having lost its meal, shoved its face through the bars of my suddenly too small cage. It shook its head back and forth. With a muffled scream, I let go of the bars and backpedaled against the current that was propelling me forward. Its long pointy snout reached about a foot into the cage, and it took all my strength to keep from crashing into it. Suddenly, the shark reared back and bit down on the bars. There was a grating clank of five rows of teeth hitting metal. My breath fired out of the respirator in panicky spurts.

As Ronald predicted, the mako was thin enough that if it tilted a bit and came at the bars on an angle, it could have fit between the eight-inch gaps. Death seemed inevitable. This was an animal that ate its own relatives, so I didn't presume it would spare me. Looking down at my turquoise and purple rubber suit, I couldn't believe I was going to leave this world dressed like a *Star Trek* character. I desperately wanted to signal for help and realized the sheer stupidity of Gus's emergency plan. Opening the top of the cage right now would be like opening your front door when a killer was trying to break into your house. The mako poked its face into the cage again. In desperation, I reached up and clung to the ceiling as the waves urged me toward its snout. Suddenly, it extracted its head and took to circling around me, eating the steady supply of chum being tossed out. I lost sight of the mako only to be blindsided as it rubbed itself alongside the bars. A couple of times, Gus dangled a fish in front of me and did the matador move, causing the shark to ram into the cage.

After about twenty minutes, someone threw out another fish, about twenty feet away from the boat. The mako darted after it, its tail whacking into the cage as it made its exit, leaving me literally and emotionally rattled. Realizing there was no more food coming, the shark lost interest and left for good. As *The Manatee* hauled me back in, I sighed a bubble trail of relief. There were high fives all around as I climbed out of the cage.

"That was so hard-core!" Gus said. "Well done!"

"So how was it?" Ronald asked.

Mostly I was just relieved to be alive, but I didn't want to disappoint him, so I mustered some enthusiasm and said dramatically, "I can still hear the sound of its teeth hitting the titanium . . ."

It was sunset when we motored into the Martha's Vineyard marina. From far away the island looked slightly foreboding with craggy cliffs and steep bluffs, but the harbor was welcoming. There were enough fishing boats to feel cozy but not crowded.

As Mandy and I changed in the cramped sleeping quarters, she leaned in conspiratorially and said, "Let's have a girls' night!"

"Sounds great!" I enthused.

Ronald and Les headed into town while Mandy and I ambled past the Victorian gingerbread-style inns that lined the main street and settled on a burger joint. *This is my kind of woman,* I thought when she kicked off dinner by asking for a "bird-bath-sized margarita." But it became increasingly clear that she was a time-release weirdo, who seems normal at first but whose freakishness unfolds over a matter of hours. At one point she launched into an extended screed against the Puerto Rican community.

"I don't mean to be racist," she began—a qualifier invariably followed by a racist statement—"but I just feel like they're trashy."

I thought, *This is coming from a woman with an aquarium on her back?*

After several drinks, her anecdotes about her boyfriend grew increasingly bizarre. It also turned out she was sleeping with Les, whom she met on a diving trip a few months before. In fact, he'd paid for her passage on this trip, but she wanted to break it off so she'd been ignoring him since they got here. Hence, her suggestion we have a girls' night out.

It was after 11:00 P.M. when we returned to the boat and accidentally stepped on Gus, who was sleeping on deck. Traffic was bad and Bill had texted that he was still hours away, so he was going to grab a hotel room and meet us in the morning. I hosed off as best I could while still wearing my swimsuit. There was no saving my hair. It looked like it was not only styled by the mice and birds from *Cinderella*, but also serving as their primary residence. I eased down the ladder into our cabin, trying not to wake Les and Ronald, who was lightly snoring. I grasped at the darkness until my hands found my cushioned bench bed. Sleeping without a blanket felt almost as vulnerable as being in the shark cage. The encrusted salt bit into my skin every time I rolled over. I dreamed my body was being attacked by millions of tiny sharks.

"Hancock!"

I was sitting cross-legged on a padded seat on deck, reading the previous day's newspaper, when I heard my name. I squinted into the morning light to see Bill climbing aboard, grinning his lopsided grin, brown curls barely contained beneath his White Sox cap.

"You're here!" I exclaimed.

He came to a stop in front of me and raised one hand in a jaunty sailor salute. "Front and centies!"

"You're in surprisingly good cheer for someone who got in at three A.M.," I said as he tossed me his backpack. "Where did you sleep?"

"In my rental car in the marina parking lot," he said with a laugh. "Didn't seem worth it to pay for a motel." He was wearing Birkenstocks, cutoff jean shorts, and a bright yellow T-shirt with JAMAICA ME CRAZY! printed across the front. That was essentially what he'd worn every day when I worked with him at the magazine, even though the office was in a corporate high-rise in Midtown Manhattan. There were introductions all around, and within minutes he was regaling the group with an anecdote about a narrowly missed ferry. They were instantly taken with him, as I knew they would be. Bill can—and will—talk to anyone.

The wind felt almost combative as we headed out of the harbor, and soon I was securing my matted hair in a ponytail to keep it out of my face. The sea was rougher than yesterday as well. The boat pitched mercilessly until the horizon resembled a possessed seesaw. Soon I was clutching the rail and throwing up again. Bill disappeared below to get me some Dramamine from his bag.

"Wow, it smells awful up here!" he said cheerfully when he returned, the smell of vomit having fully mixed with that of chum. "It smells like puke that puke puked."

An hour later, it was time to dive again. I had no desire to get in that cage again, but it *would* take care of my scary thing for another day. And the others would think it was weird if I came all this way to dive only one time.

I rooted around in the bin full of weight belts. "Is this the same weight belt I used yesterday?" I asked, holding one up. No one answered. Gus was busy helping Ronald and Mandy from the cage, and Les was fixing something on his camera and didn't look up. I shrugged and tied it around my waist. When it was time for Bill and me to climb into the cage, he offered to hold my disposable underwater camera since I was getting in first. As I was easing my body into the chilly water, I heard a plasticky clicking sound. I looked up to see Bill holding up the camera, squinting one-eyed through the viewfinder.

"Smile, you son of a bitch!" he said in his best Roy Scheider voice.

I plugged the regulator into my mouth and slipped silently beneath the surface, followed by Bill a minute later. Gus shut the top of the cage with a clang and tied it shut with the bungee cord. I could feel the cage lowering around me, but I wasn't going with it. Instead I was hovering in the middle of the cage, halfway between the ceiling and the floor. The weight belt. It must not have been the same one from yesterday, which had been heavy enough to keep me firmly on the floor. The ocean was more turbulent today. Suddenly, an enthusiastic wave pushed me upward. I bonked lightly against the rapidly descending ceiling. I shook my head to get my bearings and realized my matted hair was caught in the rubber bungee cord that held the roof shut. I was hanging from the top of the cage by my ponytail. My legs kicked out like a condemned prisoner on the gallows, fighting to the end. The weight belt was pulling me downward and the force of my hair being yanked upward lifted my mask and seawater trickled in. Trying to get leverage, I stepped on two of the cage's horizontal bars to hold myself up while I untangled my hair. As my feet stuck out over the edge, I remembered Gus telling us not to stick our hands or feet outside the cage because "sharks will take a test bite out of anything." I tore at my hair frantically. This would probably have been a good time to signal for help by opening the cage door if my hair hadn't been tied to it. After about five minutes, I ripped my ponytail free and joined Bill at the bottom of the cage, an inch of water lolling around the bottom of my mask. I tried the trick that Gus taught me to get the water out—looking up while gently pulling open the bottom of the mask—but instead more water rushed in. (Later I would find out I forgot to exhale through my nose at the same time that I'd lifted the mask.) I looked at Bill with pleading eyes. "I have water in my mask and I can't get it out! What should I do?" I wanted to shout. I pointed to my mask and he shook his head, not comprehending.

I tightened my mask, but when I pulled the strap, even more water poured in. I tried to isolate my breathing and just inhale through my mouth, but every time I automatically breathed in through my nose, too. I was sniffing seawater at a fast clip and wondering if there was enough room on a death certificate for "died of complications from unmanageable hair." By this point, the water in my mask had crept up past my nostrils. My eyes stared down the bridge of my nose like two flood victims on the roof of a house, wondering if the water was going to overtake them. If that happened, I was fucked. Not only would I be trapped in a cage, inhaling seawater, I'd be *blind* and trapped in a cage inhaling seawater. Instinct told me to swim to the surface as fast as possible, but there were sharks in the water. The question was: Did I want to drown or be eaten alive? Choose your own adventure!

I gave Bill the double thumbs-down, the international sign for "I am displeased," and motioned that I wanted to get out. I unfastened the bungee cord and opened the top of the cage, but the current was so strong that I couldn't close it again. The cage door was swinging wildly when Gus pulled us up and he couldn't grab hold of the roof. The cage slipped horizontally underneath the boat, thudding dully against the hull. It was on its side with the roof wide open so any shark could swim inside. I glanced over at Bill, who gestured helplessly. He hadn't heard Gus's speech from yesterday about needing to close the top of the cage. He had no idea what had gone wrong. I scrambled for the cage door and yanked it down but a strong wave ripped it free again, almost pulling me out of the cage. I braced the tops of my feet against the cage's horizontal bars for leverage and pulled with all my strength. As the cage jerked around, the metal tore easily into the thin skin on the tops of my feet. Then we were moving, being slowly towed forward; Gus was reeling us in. When we reached the surface, he locked the metal cage in its place at the back of the boat. I exploded out of the water, dropping my regulator and dry heaving. Hands scooped me up underneath my shoulders from behind and dragged me into the boat. "Easy, take it easy," voices urged.

I rinsed off, taking care not to aim the hose above my neck where it would surely blow the eyelashes and eyebrows right off my face. Weakly, I made my way over to an empty space on the boat deck and lay down, warm in my wetsuit in the summer sun.

"You okay over there?" Gus finally called out in a tone that implied he no longer thought me hard-core.

I nodded without opening my eyes.

"In that case, Les, you go down with Bill."

I sensed Bill was standing over me. Or, more accurately, I felt him dripping on me.

"Really, are you okay, Noelle?" Bill almost always called me Hancock; hearing him say my first name was jarring. "Do you want me to stay with you?" he asked. I was instantly moved that he'd come all this way and then offered to give up his shark dive for me.

"I'm fine. See?" I sat up, as proof of my okayness. "Now get back down there!"

He turned and made his way back to the cage, taking slow, exaggerated steps to avoid tripping. "If you have any last-minute advice on how to avoid drowning, I'm all ears," he called over his shoulder. "Seriously, my ears are *huge*!"

I rolled my eyes and smiled. "Just watch those girly curls of yours around that cable."

Eleanor didn't learn to swim until she was a mother and wanted to be able to watch over her children when they were in the water. So, in the winter of 1924, Eleanor took lessons at the YWCA pool in New York and learned to swim at the age of forty. Diving took longer—until the summer of 1939, in fact, when she was fifty-six years old and took lessons from Dorothy Dow, a junior member of her White House staff.

"Finally she could dive," Dow wrote, "not only from the side of the pool but from the diving board as well. She was anxious to perform for

the President, as he said he didn't believe she could do it. . . . So, Mrs.
R. walked out on the board, got all set in the proper form and went
in flat as could be. She could have been heard down at Poughkeepsie!
I thought the President would explode laughing, and his hand came
down on my shoulder so hard I almost fell over. Mrs. R. came up red in
the face, with a really grim expression, said nothing, walked out on the
board again, and did a perfect dive."

Les was laughing when he and Bill surfaced twenty minutes later.
"You could've gotten your hand bitten off!" he said between gasps.

Bill looked sheepish as he told us what happened. They'd been
down for a few minutes when they were greeted by an eight-foot blue
shark. Bill wanted a picture of himself giving the shark a high five.
He motioned for Les to get his camera in position; then he reached
through the bars, grabbing onto the fin. Bill barely managed to yank
his hand back into the cage as the blue turned its head and snapped at
him.

"Do you want to dive again?" Gus asked me.

"No, I'm good," I said quickly. Before I lowered my eyes, I saw the
disappointment on Gus's face.

On the ride home, everyone lapsed into that exhausted silence that
signals the end of a vacation. Bill and I sat beside each other on the
deck, knocking together companionably whenever the boat hit a big
wave. Every time I thought of myself refusing to get back in the cage, I
felt a flash of irritation. Bill's bravado only made me feel like more of a
failure. He'd also been in the cage when we were trapped underneath
the boat. But it hadn't stopped him from going back down. I'd had my
second chance, but unlike Eleanor and Bill, I hadn't tried again. I'd
had my first setback and I'd given up.

"How do you do it?" I asked. "How can you be so daring about everything?"

Bill shrugged. "I'm not so brave."

"You're just trying to make me feel better."

"I've never hit on a woman sober."

"What?"

"I'm thirty-three years old and don't have the stones to ask a woman out unless I'm drunk," he said. "So you see? We're all afraid of something."

"Except me." I chuckled. "I'm afraid of everything."

Bill's face darkened. "What's happened to you, Hancock?"

"What do you mean?"

"When you were our intern, you'd come into the office every morning and entertain us with a new story about some wild thing you'd done the night before. Like the time you were on your way home and some girl on the subway started talking smack to you? But you talked smack right back and held your ground. Even when she pulled out a knife!"

"Well, that was totally stupid."

"Where is that Noelle?" he asked impatiently. "Because I want her back. This self-deprecating shtick you've been working for the last few years is getting really old."

I was surprisingly stung by his words. I'd changed. You'd think that because I already suspected this to be true, his confirmation wouldn't be that painful to hear. But there's still hope within suspicion, a chance that your problem exists only in your imagination. To have it confirmed and articulated by someone else meant it was real.

For a long time, I stared out at the ocean. It made me think of Matt. Since he was a child Matt had spent every summer at his parents' beach house in the Hamptons, frolicking through these rough Atlantic waters. The first time Matt coaxed me into the water was also the last time. I was used to the Gulf of Mexico, where the waves don't go over two feet unless there's a hurricane. But Atlantic waves attack in a group assault, knocking down unsuspecting victims for a thorough beating. The effect is similar to being mugged. And when you finally stagger to your feet, you'll often find yourself without a bathing suit bottom. Over and over, I was knocked down, rolled, and came up sputtering.

"You have to dive under the wave," Matt instructed. "Like the surfers do."

"I *was* under the wave, Matt. I was under about eight of them simultaneously, in fact."

As I was saying this, another wave plowed into me, dragging me across a bed of crushed seashells. When I stood up, I had two bloody knees. I promptly threw my hands in the air in a leave-taking gesture.

"And that's it for me!" I told the waves. "Thanks so much! You guys have been great."

"Awww, don't leave," Matt begged.

Making my way toward shore, I called out, "I'm going to lie in the sun with the normal people who prefer to kill themselves slowly."

"He's right, you know," Dr. Bob said later when I told him the story. "The problem is in your approach, bracing yourself and trying to hold your ground where the waves are strongest. When you dive *under the wave,* it rolls over you, and you come up on the other side. Eventually you're out there happily bobbing up and down, moving with the waves instead of against them. The same thing is true for scary situations." Dr. Bob inched forward in his chair, like he was about to tell me something vital. "Rather than tensing up and trying to stand your ground when the scary situations come at you, you should dive into them. Roll with them rather than struggle against them. It's rough at first, but once you put yourself out there, it's much easier to ride the ups and downs. And it's far more enjoyable than spending your life sitting on the beach and watching."

I thought about Dr. Bob's wave metaphor as our boat pulled toward the dock. Although I was proud of my successful shark dive the day before, I hadn't achieved the same sense of accomplishment I'd had with the last big challenge. During the apex of my last swing on the trapeze, I'd experienced a joyousness that I never would've felt had I not gone up there. My shark encounters, on the other hand, had been full of terror and panic that hadn't stopped until they were over. Not all fears are worth chasing, I realized. What had I really gotten out of this? Sure, I'd *lived* and I'd have a good story to tell, but shouldn't life be about more than just survival and bragging rights? Shouldn't it be about growth? Being afraid of sharks was like being afraid of fire. There was no psychological upside to overcoming one's fear of sharks. We're supposed to be afraid of them—they're monsters! From now on I'd choose my challenges more carefully. As I stepped off the boat onto the dock, I calculated how many days I had left on the experiment: more than three hundred days. That's a lot of tomorrows.

Chapter Four

Do the things that interest you and do them with all your heart. Don't be concerned about whether people are watching you or criticizing you. The chances are that they aren't paying any attention to you.

—ELEANOR ROOSEVELT

"I just came away from the weekend wishing I was more like my friend Bill," I said wistfully.

"And what qualities does Bill have that you admire?" Dr. Bob asked.

"Well, the man has almost no fear. And he's just . . . goofy."

"When was the last time you did something truly goofy?" I opened my mouth to reply and he added, "While *not* under the influence of alcohol."

I closed my mouth and reconsidered the question. "Probably right before college. Yes, definitely, the Yale video."

"Video?" he repeated, confused.

Of all the colleges I applied to, Yale was the last to respond. I'd already been rejected from Duke and Georgetown, which sent terse letters regretting to inform me of my subpar credentials but wishing me luck at a school with lower standards. So I was bowled over when Yale wait-listed me. Immediately I unleashed an aggressive letter-writing campaign upon the admissions office. Every other day for three weeks, I mailed a letter detailing why Yale would be making a terrible mistake if I wasn't admitted. Then I got creative. I'd always loved the Dr. Seuss book *Oh, the Places You'll Go!* So I wrote my own version called *Oh, the Places I'll Go!*, rewriting the words so that the poem was about me getting into Yale. An example of one stanza:

> *There are other schools I've looked over with care,*
> *but I've made my decision: I don't wanna go there.*
> *My university? It must be the best.*
> *Positively, absolutely, it must top all the rest!*
> *A place like Yale where my thoughts can be grown,*
> *guided and nurtured, but still be my own.*
> *I won't lag behind,*
> *No, I've got the speed.*
> *Give me the chance*
> *and I'll take the lead!*

Then I acted out the poem on video. The film was shot with the help of a few friends and the special effects amounted to my mother holding a sprinkler over my head to emulate a storm system, but the trampoline sequence more than made up for it. A week after I sent in the video, I received a call from an admissions officer.

"Anyone who puts forth that kind of effort to get what she wants is clearly going places," he said. "Welcome to Yale University."

While I recounted this Dr. Bob was leaning back in his chair, laughing. "What a fan-*tas*-tic story!" he said with obvious delight. "That took balls, girl!"

I felt a twinge of jealousy for my former self, which I hadn't even realized was possible. "Yeah, I had more nerve back then."

These days I only *thought* about doing goofy things. Sometimes during serious situations—church sermons, job interviews, even sessions with Dr. Bob—I'd torture myself by imagining doing something completely stupid like standing up and shouting, "Oooga Boooga Pee Paw!" while shaking my hips and beating on my chest like King Kong. Then I'd struggle to keep a straight face and the smile out of my voice when it was my turn to talk.

Dr. Bob asked, "Where did you learn to stop being silly? When did you start taking yourself so seriously?"

"Actually I think it started at Yale. I just"—I paused and searched for the right words—"folded inward somehow. It was so intense, you know? Everyone had to be number one at everything. Students didn't just play the violin. They played Carnegie Hall at age twelve. I knew I couldn't compete, so I stopped putting myself out there."

"And after college?"

"I went to work for a newspaper. The staff prided themselves on being intellectuals. People who didn't take themselves seriously weren't taken seriously by others. Whenever I goofed off, they'd roll their eyes. So I stopped, and that part of me never really came back."

Dr. Bob nodded thoughtfully. "Goofiness is threatening to people who want to be in control of themselves all the time, who want to be serious. What stops us from acting goofy is our fear of being evaluated. But silliness can be empowering. I think you need to stop being afraid to be goofy."

I considered, right now in this moment, doing the Ooga Booga dance, but I suspected he wouldn't see this as goofiness, but as a sign I needed a referral to a neurologist. Instead I asked, "How do I do that?"

"By practicing doing goofy things." Grinning, he held his hands out to each side and I feared he was going to make jazz hands. He did. "Hey, it works for me. I'm a goofy therapist!"

The next day I signed up for a tap dancing course. In terms of goofiness, it's hard to top a group of adults wearing Mary Janes hopping around and performing elaborate routines for a nonexistent audience. One of the routines involved a phenomenally absurd knee-slapping move, then walking across the room in an exaggerated manner while waving, bringing to mind one of those cartoon frogs pumping a top hat and cane. I also put on a Santa costume left over from college and wore it around during the day. But even as I tried to distract myself with other goofy tasks, a sense of dread lurked in the pit of my stomach. There was no getting around the inevitable.

I was going to have to karaoke.

Most people start out as goofy children and grow more serious with age. Eleanor went the opposite route. One of my favorite stories about her is from a memoir by novelist Fannie Hurst. In *Anatomy of Me*, Fannie described her visit to the White House in 1933. After lunch Fannie accompanied Eleanor to the hospital, where one of her sons was recuperating after an appendectomy. Next they attended an opening of a Picasso exhibit, where Eleanor gave a speech. Then the duo returned to the White House to host a delegation of about forty educators from the Philippines and meet with an African American Baptist minister from Atlanta. After a quick change of clothes, they went out to dinner with a Roosevelt family friend. At eleven o'clock, Eleanor and Fannie returned to the White House to see the first screening of a "talking picture" from a projector that Metro-Goldwyn-Mayer had recently installed for the president. Well after midnight, Fannie crawled into bed in the Lincoln bedroom, "deciding for once to retire without even removing my makeup." Then came a knock at the door.

"Come in," Fannie said uncertainly . . .

Eleanor strode in wearing a black bathing suit, a towel draped over one arm. "Remember when I promised . . . to show you my yoga exercises?" she asked, spreading the towel on the floor. Then, to Fannie's surprise, the forty-nine-year-old First Lady "stood on her head, straight as a column, feet up in the air."

⌐

"You just have to go up there and have fun, baby!" Matt said a few days later, ushering me into a shadowy karaoke bar. Earlier in the week Matt had asked his friend Jesse—a theater critic and passionate karaoker—to recommend a place.

"Actually, I'm karaokeing this weekend with some friends I know through work," Jesse had told Matt. "They're big drama queens, but I'm sure they'd love to have you along!" Jesse's work friends, it turned out, included several real-life cabaret singers.

"I can't believe that for my return to karaoke, you brought a bunch of professional singers," I grumbled at Matt as our group settled in around a few purple velour banquettes.

"I thought he was calling them drama queens because they were high maintenance!" he said defensively. "Besides, how good can they be if they're doing karaoke?"

"How good can they be?" I repeated. "That guy over there was actually *in Cabaret!*"

Matt was one of those people who excelled at everything he tried. Usually I tried to avoid people like that, but this flaw was revealed to me slowly over the course of our relationship: The picture at his parents' house of him in high school winning the Manhattan 800-meter championship for the second year in a row. The time his college roommate asked, "Matt, what was your thesis about again? It won an award, right?" The day he took me sailing. The night we played pool and he ran the table. By the time he won a Pulitzer Prize—as part of a team of reporters, but still—I was onto him, but already in love with him. In addition to all this, he played guitar in a band and had a terrific singing voice.

Rather than argue with Matt any further, I turned my attention from him and took note of the emergency exits. It was surprisingly tony for a karaoke bar, a neon-lights-and-martinis kind of place. At least the stage wasn't very high, just a raised platform about a foot off the ground.

Dr. Bob once told me that all our fears—no matter how irrational they may seem to us today—are in some sense survival based. These impulses helped our ancestors avoid all sorts of unfortunate consequences. Claustrophobia, for instance, is related to our ancestors' vulnerability to predators while trapped in tight quarters. But in a civilized setting, the same impulses appear somewhat neurotic. In ancient times, the primitive urge to remove yourself from danger was critical to survival. In modern life, it makes you look like a flake. You duck out of an awkward social gathering with a lame excuse. If you don't feel prepared for an important exam, you fake an illness to get out of it. By obeying this instinct, you missed out on an important lesson, which was that you *did* have the ability to learn how to handle difficulty.

Chris sidled up next to me. "Plotting your escape route?" He elbowed me playfully in the ribs. He'd been at the karaoke bar that night when I'd fled in terror. Next to him, his boyfriend, Cub, grinned broadly, revealing two rows of dimples. As always, I had to make an effort to keep my eyes from lingering on Cub's face. Chris is very attractive with his delicate features and lanky body. But Cub has one of those athletic bodies and wholesome, handsome faces that made you proud to be an American. They are the best-looking and best-dressed couple I know. Their chic, preppy style is so similar they've been known to show up at parties inadvertently wearing the same outfit.

"Oh, you came!" I squealed, throwing my arms around them. Just the sight of them made me less tense. The cocktail waitress arrived with their Bombay Sapphire and tonics, and Chris nudged me. "Care to fortify yourself with a beverage, Mariah?"

"I have to stay sharp for my fans," I deadpanned as they clinked their glasses together to toast. Dr. Bob also said that if you use alcohol to shield yourself from embarrassment or criticism, you're not really facing the fear. And the only way to get over a fear is to feel it. Which was why, against my better judgment, I was going up there sober.

"Just go after Cub and me," Chris suggested. "We're so terrible that we'll make you look good."

I flipped through the song list book. A ballad? No, something more goofy, but not trying-too-hard goofy. I was overthinking this. Finally I picked a song that I suspected would invite ridicule, wrote down the number code on a slip of paper, and followed Chris to the karaoke machine.

"If you punch in your code after me, your song should come up right after ours." He picked up a remote control, pressed some buttons. "There are some songs in the backlog, though. It'll be at least twenty minutes before your turn."

The first member of our group, a meek-looking fellow named Michael, did a booming rendition of "Some People" from *Gypsy*. It was over four minutes long, but he held the audience's attention the entire time. Another guy went and did something from *Sweeney Todd* that I didn't recognize, but he sang it well.

Matt patted my arm reassuringly. "Don't worry. I'll go put in something less . . . stirring. Mix it up a bit." He left but came back a minute later looking sheepish.

"Hey, what's your song code? I think I pressed the wrong button and accidentally erased a few songs."

Without taking my eyes off the singer, I rummaged in my back pocket until I found the crumpled slip of paper and handed it over. He returned and we watched a trio of NYU coeds giggle along to Madonna. When Chris and Cub's song came on, I couldn't help but smile. They performed a duet of Meat Loaf's "I Would Do Anything for Love," and Chris had been right—they were monumentally and wonderfully bad. And I loved them for it. When they eked out the last off-key note, I gathered myself and stood up. Might as well get this over with.

Matt slapped me on the ass. "Go get 'em!"

But then—something was wrong. Flashing across the screen was the title "Creep" by Radiohead. This wasn't my song.

"This is *my* song!" Matt said, standing up in surprise. "I'm sorry, honey. When I erased your code and put it back in, it must have logged it in after mine." He shrugged apologetically and hurried up to the stage to take the mic from Chris and Cub, who looked confused.

"What's going on?" Chris asked, sliding into my banquette with Cub. "I thought you were going after us."

"Matt accidentally swapped the order of our songs," I said uneasily. "He's going before me, not after me."

Matt stepped into the spotlight and smiled smoothly at the audience. A natural performer. "How are you folks doing tonight? Listen, I'm going to need some backup on this one. Can I get a volunteer to come help me out up here?"

"I'll do it!" The guy from *Cabaret* charged to the front where he greedily snatched up the second mic. The first moans of guitar filled the room. Any hopes I had of Matt dialing it back for my sake were dashed as he crooned soulfully into the mic, instantly quieting the room. *Cabaret* guy kicked in a few moments later with a truly beautiful tenor. They were incredible.

"I can't believe it!" I turned to Chris, aghast. "It's happening again! I have to follow a showstopper!"

"And he brought *a backup gay*!" Chris gasped. "That ain't right."

I spent the rest of Matt's song trying not to think bad thoughts about him. And then . . . thunderous applause. My turn. I didn't look at Matt as he handed the microphone over. I was scared of what I'd do with the mic if I did. Instead, through gritted teeth, I muttered, "Thanks, babe."

For my karaoke debut I'd chosen Salt-N-Pepa's "Shoop," a rap song from my formative years. I looked out at the audience without really seeing them. When the music began, the drama queens gave a low cheer of approval, bopping in their seats to the backbeat.

Taking a deep breath, I recited:

"Here I go! Here I go! Here I go again! Girls, what's my weakness?"

"Men!" the drama queens answered.

"Okay then," I continued, *"chillin' chillin', mindin' my bidness. Yo Salt, I looked around and I couldn't believe this . . ."*

So overpowering was the music that I couldn't hear my voice coming through the speakers. I couldn't even hear the words coming out of my mouth, which threw me at first. But soon I was totally invested, doing both parts of the duet myself, seamless in my transitions, changing the octave of my voice to indicate whether we were hearing from Salt or Pepa. I was doing some moves, too, holding my free arm straight out in front of me, slicing through the air as the hip-hop artists do. A little hip gyration action. I was taking it down to the floor! I was owning this! I even sang the part where the random guy interjects to give his ringing endorsement of fellatio. Oh wait—the karaoke version skipped over the guy's line! Censored it. Now I was way behind on the lyrics! I'd overreached! I stopped the dancing shenanigans and studied the teleprompter intently, stumbling through the words, trying to regain my vocal footing.

I could hear myself now. It was my voice but not my voice. Too thin, watered down. The same one I heard every time I transcribed one of my tape-recorded celebrity interviews, cringing at my voice, high pitched and nervously asking questions. From the audience, Chris's familiar laugh rang out above the music. It cut through my anxiety. The ridiculousness of what I was doing became clear, and the most amazing thing happened: I stopped caring. Just like that.

I caught up—gloriously!—on the chorus. *"Shoop shoop-be-doop, shoop-be-doop, shoop-be-doop-be-doop-be-doop,"* I sang. *"Baby baaaaby! Don't you know, I want to shoop baby!"*

Then the song was over, sooner than I expected. With a sheepish grin, I handed off the microphone. There was a smattering of applause from the rest of the audience. The drama queens, sensing the momentousness of the occasion, hooted and hollered as I made my way back to my seat. The guy from *Cabaret* intercepted me with a hand on my shoulder, saying he respected me for "fully committing to a vision and going with it," which may or may not have been a compliment. No matter. I felt liberated in some small way, having confronted this moment I'd been sidestepping for so long.

"Nice, babe," Matt said as I settled in next to him. "I think you could've gone bigger with it, though. Less head voice, more projection."

"What?"

"I'm just saying don't be afraid to come from the diaphragm, you know?"

"The diaphragm," I repeated.

"You just sounded a little pitchy, that's all. Maybe focus less on the dancing next time? I think it threw you off your game and, to be honest, it was a little distracting."

I stared at him.

"What?"

"You're giving me *notes*? On my karaoke rap song? And after what you did to me up there?"

"I'm just trying to be helpful." He looked genuinely confused. "What did I do?"

"Bringing down the house like that with fucking Tommy Tune on background vocals—"

"He volunteered!"

"After you *asked* for a volunteer! It's like no matter what we do, you always end up being the star."

"Well, I don't *mean* to be."

"I know! That's the worst part!" I let my head fall back against the wall. "Sometimes I just wish you were a little less . . . perfect." But I said it with a smile, willing away my irritation so it didn't ruin the mood.

He leaned over and kissed me on the lips. "Sorry, honey, I'll try to be better from now on. Or worse, rather." He turned back to the stage, but I continued to watch him.

The only thing Matt didn't excel at was being impressed. It took a lot to wow him. He was supportive, which was not exactly the same thing as admiring; he was also critical. I'd always thought that a couple should be, to put it simply, big fans of each other. But what happened when one person was more of a fan? I wondered, not for the first time, if Matt and I were a bad match. He was the prizewinning reporter and I larked around writing fluff. I creaked out ridiculous '90s rap songs while he brought an audience to its feet with a soulful ballad. My accomplishments would always look duller next to his, my faults more glaring. Would being married to him be like attending Yale for the rest of my life? Would I always feel like I was struggling to keep up and didn't quite deserve to be there? And would he eventually grow weary of my slowing him down?

Chapter Five

⁓

The giving of love is an education in itself.

—ELEANOR ROOSEVELT

Franklin was strolling down the aisle to stretch his legs when he spotted Eleanor sitting alone on the train. It was a muggy summer day in 1902, and he and his mother, Sara, were en route to his family's estate in Hyde Park, New York. Franklin and Eleanor were fifth cousins, once removed, and barely knew each other; but he made his way over, and soon the two were chatting amiably. She was eighteen at the time and on her way to Grandmother Hall's summer house. He invited Eleanor to join them in their car. Sara Roosevelt greeted Eleanor coolly—setting the tone for the forty years that would follow.

By all accounts, Eleanor and Franklin were complete opposites. He was handsome, free spirited, humorous; she was serious and insecure about her appearance. On New Year's Day they ran into each other again at their uncle Theodore's annual reception at the White House. From then on, the earnest debutante and the jovial Harvard man saw each other often. Eleanor attended Franklin's twenty-first birthday party later that month. He invited her to weekend house parties at Hyde Park, and to his mother's summer cottage on Campobello Island off the coast of Maine. The two of them, of course, were never allowed in each other's company without a chaperone.

Eleanor later described some of the strict rules that governed meetings between single men and women at that time: "It was understood that no girl was interested in a man or showed any liking for him until he had made all the advances." She added that "the idea that you would permit any man to kiss you before you were engaged to him never even crossed my mind."

Over a hundred years later, I was at a magazine party chatting with friends, thinking it was late and I should go home, when a handsome guy with thick dark brown hair strode confidently into the room in full black-tie regalia. Any other guy would've looked ridiculous wearing a tuxedo in a room full of jeans, but somehow he just made everyone else look underdressed. I'd seen him around. I knew his name was Matt; he was an up-and-coming reporter and had a reputation as a player.

This was back when I was still socializing, before my job as a blogger took over my life and I stopped going out, and three and a half years before I lost that job. If I saw Matt at a party now—well, I probably wouldn't have gone to the party in the first place, but I certainly wouldn't do what I did next: I waited until he'd secured a drink, and as he was walking away from the bar I fell in step next to him.

"Scotch straight up, huh?" I gestured to the glass in his hand. "You know what goes great with that?"

"What?"

"Me." Without waiting for a response, I pulled the drink from his hand and took a long swig. Up close he looked to be in his late twenties but already had laugh lines.

His eyes danced with amusement as I handed him back his glass, but otherwise his expression remained unchanged. The man was smooth. His eyes, by the way, are periwinkle. He is one of those people who would've been incredibly attractive anyway, but God went ahead and threw in periwinkle eyes in case you didn't get the point.

"Thanks for the drink." I started to walk away.

"Hey, you've gotta pay for that, you know," he called after me in a teasing voice. *Got him.* I whirled around.

"Oh please," I scoffed. "You weren't going to drink all that." I gave him an exaggerated once-over and declared, "You look like a lightweight to me."

"For all you know, I've already had five of these."

I planted a hand on one hip. "I know for a fact you haven't."

"The only way you could know that is if you'd been checking me out since I got here."

We grinned at each other, feeling like stars in a 1940s screwball comedy. Two hours and several scotches later I'd learned he was raised in Manhattan and graduated from Princeton.

"Princeton, huh?" I tried not to look impressed. "Tell me, Princeton, do you always go around wearing tuxedos to casual parties?"

He smiled. "Only occasionally."

"Yeah, I almost rocked my tux tonight but opted for the rugged outlaw look instead, as you can see." I gestured to my cardigan and tweed slacks.

He laughed. "I was covering a black-tie political dinner for the paper earlier. It's been a *long* day. In fact, I should probably be heading home." He looked down at his watch. Then his eyes flicked up toward me. "Would you like to come along?"

I would. But I also guessed that he was a guy to whom things came easily, and things that came easily were often easily discarded. So I fixed him with an arch look and said, "Oh, I haven't had nearly enough scotch for that."

He leaned in and I could feel the heat from our bodies mingling as he said softly, "That's a shame." He handed me his scotch glass with a wink. "At least I know you can give this a good home. It was nice meeting you, Noelle."

I waited three days before looking him up on his newspaper's website and sending him an e-mail.

"I knew you were going to e-mail me," he wrote back. His brazenness was both annoying and enormously appealing.

The rules of courtship had relaxed somewhat in the hundred years since Franklin was wooing Eleanor. Matt and I dated seriously over the next few months and officially became a couple on New Year's Eve. The bathroom line at our party was horrendously long, so he stood guard while I crouched on a fire escape in my satin cocktail dress and peed into a plastic goblet. I knew right then he was the pick of the litter.

"So what's your plan for this weekend?" Dr. Bob asked.

"I think we've progressed beyond small talk, don't you?"

"With your fear conquering, I mean. What's on the list?"

"I'm taking my fears on the road." I leaned back and linked my hands behind my head. "Matt and I are going to a wedding on Nantucket Island this weekend."

"That sounds terrifying," he joked.

"Believe me, it is. Instead of driving with him, I'm taking a plane there and back because I'm afraid of flying."

Dr. Bob looked unconvinced. "That sounds more like someone looking for a way to get out of a long car ride."

"This isn't just any plane. It's one of those hideous puddle jumper planes that fall out of the sky on a remarkably regular basis. I always swore I'd never fly on one of those."

He nodded his head from side to side wearing an expression that said "not bad." "And what about Saturday?"

"I don't know what I'm going to do on Saturday. It's hard to plan a fear for a place you've never been. I'll have to see what comes up."

"This couple getting married—are they Matt's friends or yours?"

"Please!" I laughed. "My friends don't have weddings. They have Xboxes."

Matt's friends were four years older and in a completely different phase of life. They owned houses and garlic presses. They knew where to put the zeros and the ones on their W-4 forms without asking their parents. My friends still stopped to examine the furniture New Yorkers put out on the street for trash and pondered whether or not they could carry it twenty blocks back to their apartments.

Dr. Bob propped his dimpled chin on one fist. "Do you and Matt ever talk about getting married?"

"Never."

"Do you think about marriage?"

Sure I'd thought about it, but in the way that a middle schooler thinks about college. Marriage had always felt inevitable but so far in the future that it didn't seem real. I couldn't even *picture* myself married. I'd look ridiculous wearing a diamond. I still wore T-shirts featuring dinosaurs. If I got married, I'd have to upgrade my entire wardrobe to match one finger.

"I guess I always had it in my head I wouldn't get married until I was in my thirties."

My parents had eloped when my mom was twenty-two and my dad was twenty-four. By the time my mom was thirty, she had a two-year-old and a five-year-old. Every night my mom sat on the back porch for hours, smoking cigarettes and reading romance novels until the Texas humidity chased her back indoors. Sometimes I would sit with her, watching moths being lured to their deaths by our bug zapper. A few times over the years, my mom would look up from her book, sigh a cloud of smoke toward the tin porch roof, and say, "Promise me you won't get married until you're at least thirty."

"You're almost thirty," Dr. Bob pointed out. "Do you want to marry Matt?"

I knew this was therapy, but the question struck me as a little personal. "I mean, I don't know," I stammered. "Maybe. You know?"

Dr. Bob waited it out. When I'd been quiet for a while, he prompted, "But you love him, right?"

"Of course! Definitely. It's just . . . I just thought I would *know* when I met The One. But with Matt I guess I don't know without a doubt. Which is ridiculous, right? He's sexy and smart and caring and patient—can you believe that in three years I've never heard him raise his voice?—and hardworking and honest. If you overlook the fact that he has finger toes, he's the ideal man."

"Finger toes?"

"His toes. They're really long and look like fingers. I make fun of them all the time."

"Really?" Dr. Bob looked a little disturbed. "I've never seen a finger toe."

"Well, consider yourself lucky. They're terrible. My point is, Matt is basically the perfect guy, but are we one soul dwelling in two different bodies? I wouldn't go that far. How do I know if he's perfect *for me*?"

"You don't." Dr. Bob crossed his ankle over his knee and adjusted his khakis.

"But aren't you supposed to know these things for sure? Maybe the fact that I don't know whether Matt is The One is a sign that he *isn't* The One?" My eyes fell on Dr. Bob's gold wedding band. Twenty years ago he married a Rockette, and they'd been happily wed since.

"Have you thought about bringing your concerns up to Matt?"

"No. Things are perfect between us right now. Why would I mess with that?"

He waited for me to go on.

"And what if I marry Matt, then later I meet some guy who's exactly as perfect as Matt, but he's also hysterically funny and not allergic to cats?"

"Matt's not funny?"

"Sure, he's normal person funny, but what if the guy I meet is, like, Conan O'Brien funny? What then?"

Dr. Bob made a pyramid with his fingers. "Remember when we talked about perfectionism?"

"I remember."

Perfectionism is the fear of making mistakes. There are two sides to perfectionism. At its best, it's motivating and inspires you to set high goals for yourself. But it can also get out of control. Perfectionists can turn into workaholics because their efforts never feel good enough. They engage in all-or-nothing thinking about their performance—if it isn't perfect, it's horrible. They give up easily. They procrastinate on goals, waiting for inspiration to strike or the timing to feel right. They avoid social situations if they aren't feeling "on." They organize their lives around avoiding mistakes and end up missing wonderful opportunities.

Dr. Bob leaned back in his chair and replaced the cap on his pen. "Here's the reality of life," he said. "You make decisions with imperfect information and achieve imperfect results. The alternative is to never make a decision and never achieve results. There's no guarantee you or your spouse won't get bored and find someone else. But taking your chances with someone who is nearly right is a better bet than waiting for a perfect partner."

The next day I was standing on the runway at the airport gazing uneasily at the prop plane. It was about the size of my parents' SUV and seated six people. On top of that, we were about to fly into a rainstorm. I pulled out my cell phone and called Matt. He was driving down from Albany and would take the ferry to Nantucket, where he'd meet me at our B&B.

"Remember the last scene in *La Bamba* right before Ritchie Valens gets on that tiny-ass plane and flies into a snowstorm and *dies*?"

"Vaguely."

"That's what this feels like right now," I said, "except I haven't accomplished anything noteworthy yet."

"At least you've found your scary thing for the day. Any last words?"

Remembering yesterday's conversation with Dr. Bob, I joked, "Yeah, if the plane goes down, you're not allowed to move on and find someone else. If you do, I'll haunt you and your wife until you divorce her and take a vow of celibacy." Matt laughed, a little too heartily for my taste.

Five minutes later the puddle jumper was fighting its way through the air, the rain dribbling like beads of sweat down the windows. Flying made me rethink my life and consider my death like no other form of transportation. You hear about those crashes where the aircraft goes down with such zeal that the wreckage is reduced to pieces no bigger than a Post-it note. Maybe skydiving enthusiasts could find the silver lining and enjoy the free fall, but I hated the sensation of losing my stomach. During childhood trips to Six Flags, when my friends went on roller coasters, I sat on a bench and guarded everyone's purses.

Each time the plane bobbled, I inhaled sharply and tensed up. My position of choice during turbulence was to clench the edges of the seat cushion and pull upward. The lone upside to a dramatic death was, of course, the prospect of making the cover of the *New York Post*. But now, as I tried to conjure some punchy headlines, I worried that my passing wouldn't have as much significance because I wasn't engaged. If Matt and I were betrothed, I could make the front page: "Fiancé Recalls Last Phone Call with Plane Crash Victim: 'Never Marry.'" But as a free agent I didn't stand a chance. Sure, it was sad when someone's girlfriend died in a plane crash, but when a *fiancée* died in a plane crash? Then you're cooking with propane. Now there would be no wrenching scene with investigators handing over a half-scorched engagement ring, asking, "Sir, was she wearing this the last time you saw her?" and Matt collapsing into their arms, sobbing, while they fanned his face. No, instead they'd have to go the old "identifying body marks" route. They'd take

Matt aside and ask, sotto voce, "Sir, did your girlfriend have a tattoo of a dolphin on her butt?" (It seemed like a good idea at the time. But what doesn't when you're sixteen?)

The plane dipped abruptly, and the passengers collectively gasped. "Sorry, folks! Little bit of rough air here," the pilot called out from the "cockpit," which had no door and was so close he didn't have to raise his voice. I thought about who would come to my funeral. I was anxious that my media friends wouldn't have anything to talk about with my college friends. Then I realized they'd be talking about me so it wouldn't matter.

"To tell you the truth, I hadn't seen her that much in the last year," someone would say over cheese cubes speared with festively colored foil toothpicks.

"Me neither," another would jump in. "I did get a kindly worded text message on my birthday, though."

After fifty minutes of the plane never really calming down, it began its final descent and we wobbled down to the sweet, wet earth.

As I staggered off the plane, all I wanted was to get to our B&B. Everything about the island was quaint—the gray shingled cottages, cobblestone streets, even the plump raindrops. When I arrived, our small, surprisingly airy room was empty; Matt must have been caught in traffic. I wriggled out of my damp T-shirt and waterlogged jeans, kicking them across the wooden floor in the general direction of the radiator. Wearing only my bra and panties, I vaulted onto the high four-poster bed, sprawling on my stomach on the white down comforter.

Television options in the late afternoon were grim. I settled on a romantic comedy about a spoiled figure skater forced to team up with an oafish ex–hockey player to compete at the Olympics. Romantic comedies, I'd noticed, occasionally ended with a wedding, but they were almost never about marriage. Movies about marriage were dramas. Things ended badly for married couples in movies. Jack Nicholson tried to break into the bathroom with an axe. Glenn Close was shot in the bathtub. Thelma escaped an abusive husband only to drive off a cliff with Louise. Matt burst through the door of our hotel room just as the figure skater beaned the hockey player in the head with a puck.

He took in my bra and panties ensemble. "No need to get all dressed up for me. But I'm glad you did!"

I laughed. "It was all part of my master plan. Now get over here so I can have my way with you."

He set down his leather duffel, hung up his trench coat, and pulled off his wet shoes and socks. In three strides he was standing over me, leaning down to give me a kiss. Right before our lips touched, he shook the rain out of his hair, making me squeal. He flopped down on the bed next to me and propped himself up on one elbow. I wiggled my eyebrows suggestively and set upon unbuttoning his shirt.

"Let's get those wet clothes off," I said. As he turned his face in to mine, his feet grazed my ankle and I stopped. "But first you have to get those finger toes off me."

He grinned and pressed his feet more firmly into my ankle.

I cringed. "Seriously, I love you a little less right now."

"Oh, shut up and kiss me."

The next day, the ceremony was all whiteness and light. The inside of the church was painted a gleaming eggshell and the high arched windows bathed the room in sunlight. Even the bride and groom were pale blonds. While Tom and Casey vowed to spend the rest of their lives together, I shifted uncomfortably in our pew. Just as flying forces you to reevaluate your life, weddings force you to reevaluate your relationship. Would Matt and I be repeating those lines to each other one day? Or would we be saying them to someone else whom we hadn't even met yet? That thought filled me with sadness and I squeezed Matt's hand, as if to confirm that he was still there.

At the reception I was introduced to Matt's ex-girlfriend of five years and her fiancé. Their wedding was in two weeks. The fiancé and I shook hands and tried not to picture our significant others having sex.

After they walked away, I asked Matt, "Are you okay? Was that weird at all, seeing her?"

"Nah, we broke up eight years ago." Noticing that I looked anxious he added, "Nothing to worry about."

I smiled at him. "I know. It's not that. I haven't done my scary thing for the day yet," I said. "Usually something would've come up by now, but it hasn't." I cast a desperate glance around the room, looking for some daunting situation I could throw myself into.

He thought for a moment. A slow, mischievous smile spread across his face. "I have an idea. Follow me."

Before I could protest, he had placed my hand in the crook of his elbow in an exaggerated gentlemanly gesture and was leading me up the stairs. When we reached the end of the hallway, he opened a door and steered me into a room. Suddenly we were standing on fluffy peach carpeting, the kind that begged to be walked on in bare feet. There was a large canopy bed in the center of the room.

Matt moved behind me. "Now," he said, sliding his hands over my bare shoulders, "let's get these wet clothes off."

"But my clothes aren't wet."

He grinned naughtily. "Let's get them off anyway."

"Wait! We can't defile Casey's bridal suite!" I hissed, untangling myself from his grasp. "It's disrespectful and not entirely sanitary."

"This isn't Casey's suite. Her room is at the other side of the hotel. There's another wedding going on here tonight. But don't worry," he murmured into my ear, "everyone's going to be downstairs for *hours*."

"Still, it's not right," I insisted. "Come on, let's go."

As I started to walk away, Matt caught my hand and twirled me into the deluxe bathroom, agleam in ivory marble. Pressing his body into mine, he backed me against the door, which eased shut with a barely audible click. He skimmed his fingers across my hip and flicked the lock.

I wasn't one of those people who got a thrill from having sex in non-sex-having places. The most daring place I'd had sex was our shower in Aruba, which wasn't really adventurous but for the off chance we might slip, crack our heads open, and have our pruney, fused-together dead bodies discovered by the turndown maid. This wasn't much more courageous. Did having sex in someone else's hotel room make me nervous? Yes? Was it technically a fear? Under normal circumstances, it probably wouldn't pass inspection. But what the hell? I'd wave it through. Barring any forced participation in a dance floor conga line, this might be my last scary moment of the evening.

"There's nothing sacred about a bathroom though, right?" Matt said. His lips were already working their way down my neck.

"There won't be when we're done with it." I laughed, letting him pull me away from the door. As he kissed me, there was a soft whirring sound as he unzipped my navy satin dress. I closed my eyes and relaxed into him.

"What's that?" I whispered into his lips.

"What?"

We waited, and out of the silence came the unmistakable rattle of the doorknob.

"Why is this door locked?" a high-pitched voice screeched.

"Oh my God!" I mouthed at Matt. Like a dog chasing its tail, I spun wildly trying to reach the zipper at the back of my dress. I hoisted it up so fast that I pinched my back fat. I screamed silently.

Matt scanned the room for a place to hide. No shower curtain or linen closet.

A chorus of female voices attempted to soothe the woman's nerves. Bridesmaids.

"It wasn't locked earlier!"

"Are you sure it's locked? Maybe it's just stuck."

"There has to be a key somewhere. I'm sure that old lady who checked us in has one."

"Well, let's *find her*," the screechy voice ordered. "I'm not using the communal bathroom at my own fucking wedding!"

I pressed my ear to the door. The wood was cool and smelled faintly of chemicals. I heard her dress swooshing indignantly as she turned to walk away, and the slightly frantic steps of the bridesmaids following her down the hall. When the sounds of bustling taffeta had faded into the distance, I whispered to Matt, "Okay, I think they're gone. Let's sneak out while they're searching for—"

Before I could finish, the voices were back and growing louder.

"It shouldn't be locked," trilled a matronly voice. "I assure you, it's your own private bridal suite. No one else is permitted to use it."

Matt looked hopefully toward the window. A two-story drop. I snatched up a couple of fluffy peach bath towels. Maybe we could knot them together and rappel out the window like in the cartoons? No, there was no time. The voices had stopped outside the door. "I believe this is the correct one," the hotel matron was saying over the sound of tinkling keys.

Screw the towels. I dropped them to the floor. We stared at each other in frozen horror.

"You can hold on to this key for the rest of the night," the woman reassured the bride. There was a sound of metal hitting metal as the key entered the lock.

For a moment I considered hiding behind Matt, who would surely come up with some charming excuse when that door opened. Instead, I stood a little taller and smoothed down my dress. We could wait for them to storm in, I realized, or we could go down like heroes. I looked over at Matt and he nodded. Then I reached over and flung open the door with a flourish. A sixtysomething woman fiddling with the lock jumped back. Next to her was a bride with jet-black hair teased high on her head. Her hands, which had been on her hips, instinctively sprung into the air, acrylic nails ready to attack. Three bridesmaids in strapless lavender gowns squeaked in surprise. Matt and I linked arms. With straight faces and heads held high, we marched out of the bathroom. As we squeezed past the bride, the situation finally registered. Her forehead wrinkled in disgust.

"ARE YOU KIDDING ME?" she shrieked, nostrils flaring. "WHAT THE FU—"

"Run!" I whispered to Matt and we charged, giggling, down the stairs.

The majority of guests were from Ireland, including the guy sitting next to us at dinner. We'd just sat down but he was already in that happy, drunk place where you can no longer tell what's inappropriate and you no longer care. While tuxedoed waiters passed out the salads, he turned to Matt and me.

"So are you two getting married then?" he asked loudly in his booming Irish brogue.

There it was. The question we hadn't raised in three years of dating. It had only taken a total stranger and about half a bottle of Jameson to bring it up. All heads swiveled in our direction, and I turned my head as well and stuffed a dinner roll in my mouth. When I turned back my cheeks were engorged with French bread, so everyone's attention shifted to Matt. He paused. I wondered what he was going to say.

After nine months of dating, Franklin proposed to Eleanor during the weekend of the Harvard-Yale football game. He had invited her as his guest and on November 23, 1902, they managed to ditch their chaperones and slip away for a walk by themselves. By the time they returned, he'd asked her to marry him and she'd said yes.

Predictably, Franklin's mother disapproved of the match. She'd wanted a more attractive wife for her only son. In a move that's almost impressive in its deviousness, she suggested the couple keep their engagement secret for a year. Then she took Franklin on a five-week Caribbean cruise, hoping he'd lose interest in Eleanor. Their feelings only grew stronger. Franklin returned for his final term at Harvard, and they wrote passionate love letters back and forth.

"You are never out of my thoughts dear for one moment," Eleanor wrote to her fiancé. "Everything is changed for me now. I am so happy. Oh! So happy & I love you so dearly." When Eleanor told her grandmother about the proposal, Mrs. Hall asked if she was really in love. "I solemnly answered 'yes,'" Eleanor later said, "and yet I know now that it was years later before I understood what being in love or what loving really meant."

They married on St. Patrick's Day in 1905. Uncle Teddy escorted Eleanor down the aisle, having chosen this date because he'd be in town to kick off the St. Patrick's Day parade. Eleanor was, in her words, "decked out beyond description." Her dress was constructed of stiff satin and the same lace her mother and grandmother wore at their weddings. She pinned her veil with a diamond crescent that belonged to her mother. Sara, who was never one for subtlety, gave Eleanor a high-necked pearl dog collar with diamond bars. It was from Tiffany and cost $4,000, but the symbolism was priceless.

Eleanor was only twenty years old and Franklin was twenty-three. Of course, the upside to marrying your cousin is you don't have to change your name. After the ceremony, the president quipped, "Well, Franklin, there's nothing like keeping the name in the family."

Eleanor's mentor from Allenswood Academy, Madame Souvestre, was battling cancer and unable to attend. Instead she sent a telegram that had only one word: *Happiness.* She died two weeks later.

Matt still hadn't answered the question, and a nervous tension settled over the table. He looked decidedly uncomfortable.

"Well, it's just—" Before he could finish, a tuxedoed waiter appeared with salads for everyone. Small talk resumed. Matt visibly exhaled with relief. *It's just what?* This whole time I'd been fretting over whether Matt was The One for me, but I'd been actively suppressing my fear that he might be having doubts about *me*. The song "Proud Mary" ended and the band moved on to "The Way You Look Tonight," one of my favorites.

Matt nudged me. "Let's dance."

The shuttle back to our B&B was packed with drunk Irishmen yelling college fight songs. Matt's ex-girlfriend was in the car, too, chattering away, but Matt and I were quiet.

We almost never fought, but when we were back in our room, he got annoyed at me for using too much of his saline solution and I snapped at him for BlackBerrying in bed. Before I'd even finished brushing my teeth, he reached for the lamp on the bedside table and flicked it off with irritation.

Our mood was still subdued when he kissed me good-bye the next morning to catch an early ferry. Two hours later I strapped myself into my second undersize plane of the weekend. As the wheels lifted off, I thought about the long pause Matt took after the question *Are you two getting married?* That drunken Irishman had given me an opening, and I should have taken it. Perhaps not right then over the salad course, but maybe later in our hotel room. I couldn't have planned a better opportunity to face my fear of The Marriage Talk. That was the fear I should have tackled yesterday, not hijacking a bridal suite for a quickie. But I was too afraid of what Matt was going to say. I was afraid of that pause and what it meant. All of a sudden I missed Matt terribly. I looked out the window and watched until we were so far away that the island was no longer charming, just a brown lump in the sea, ugly and unrecognizable.

Chapter Six

~

The encouraging thing is that every time you meet
a situation, though you may think at the time it is
an impossibility and you go through the tortures
of the damned, once you have met it and lived
through it you find that forever after you are freer
than you ever were before. . . . You gain strength,
courage, and confidence by every experience in
which you really stop to look fear in the face. You
are able to say to yourself, "I lived through this
horror. I can take the next thing that comes along."

—ELEANOR ROOSEVELT

At the end of the day, I still hated flying. The flight to Nantucket
had done nothing to quell my fear of air travel. By the time
our little plane touched down on the mainland, I knew I had
to learn to fly. Just being a passenger on a flight hadn't felt like enough
to fully face my fear. I needed to be in the pilot's seat, to see what it
felt like to be in charge of my own fate while tens of thousands of feet
above ground.

Another reason to give flying its due was because Eleanor had
adored it. She flew more passenger miles in the 1920s and 1930s than

any woman in the world, prompting *Good Housekeeping* to dub her "our flying First Lady." It was harder for Franklin to travel because of his paralysis, so she flew all over the world on his behalf, at a time when most Americans still considered air travel unsafe. She flew on a government-issued C-87A aircraft called *Guess Where II.* It was originally intended for Franklin until it came to light that C-87As had a weakness for crashing and catching on fire. So the Secret Service approved it for Eleanor's use instead.

In April 1933, Amelia Earhart attended a black-tie dinner at the White House. Eleanor had never flown at night and listened, enraptured, as Amelia described what it was like to look down and see all of Washington, D.C., twinkling below you. On a lark, Amelia suggested that she and Eleanor fly to Baltimore and back that very night. Within an hour the women were airborne, still wearing their evening dresses, gloves, and heels. Even I, hater of flying, was charmed by this story—two plucky dames ditching their boring dinner party to go joyriding in formal wear. The press had gathered at the airport by the time they landed.

"How do you feel being piloted by a woman?" a reporter asked Eleanor.

"Absolutely safe," she replied. "I'd give a lot to do it myself." Amelia offered to give Eleanor lessons, and the First Lady went so far as to receive her student pilot's license, but Franklin discouraged her from pursuing it.

"I know how Eleanor drives a car," he reportedly said. "Imagine her flying an airplane."

If I could get my pilot's license, I could finish what Eleanor started! But getting a pilot's license, I soon discovered, cost between $7,000 and $10,000 and required a minimum of forty hours of flight training with an instructor. There had to be another way. Then I remembered a conversation I'd had with an investment banker friend at a party a few years ago. He'd told me that he'd had a few too many at the company banquet and bid on fighter pilot lessons in a silent auction. He and a friend had gone dogfighting earlier that afternoon.

"It's a civilian dogfighting school where regular people get to play fighter pilot for a day," he'd explained.

"Like a computer simulation?" I'd asked hopefully.

"Naw, dude, you're actually up in a plane at five thousand feet. The instructor is there to take off and land, but otherwise you're flying that motherfucker! We were doing all these crazy-ass tricks!" He mimed flying an airplane, spilling a bit of his drink. "At one point I almost blacked out from the g-forces! It was awesome!"

It sounded awful, which was exactly why I felt compelled to do it. I e-mailed him and got the phone number of the company, Air Combat USA. When I called, I found out that the next available opening was a month away. It was no bargain, but it was certainly more affordable than getting a pilot's license. I gave them my credit card number before I could came to my senses. The operator took down my information, explaining the strict financial penalty they would unleash upon my MasterCard should I cancel.

The next weeks were filled with dread. Fighter piloting loomed in the distance, ugly and terrible, above all the other challenges. When the flying lesson was days away, I thumbed, warily, through the packet of information Air Combat sent. I read the letter of instruction:

Your mission is scheduled to begin at 1300 hours. Please report for duty 15 minutes early to allow time to suit up in your flight suit and pick out your helmet and parachute.

At first I found the military talk sort of endearing. Then I reread the second sentence. Wait, are they letting me pick out my parachute? That's putting a little too much trust in your customer, isn't it? I don't even think they let you pick your own lobster out of the tank at Red Lobster anymore. There was a promotional DVD, which I popped into my computer. The narrator had a voice for infomercials, the ones selling powerful cleaning agents. The video opened at an airport with shots of extraordinarily ordinary-looking people arriving on the scene wearing jean shorts and T-shirts with the sleeves ripped off. Then they were zipping themselves into jumpsuits and climbing into planes with Air Combat instructors. In his loud-but-not-loud voice, the narrator said: *"You're probably thinking, 'Wait a second! I've never done this before! I don't even know how to fly!' That's the beauty of what we do. No experience is necessary. You just show up and we handle the rest."*

The planes were taking off and we'd moved on to the flying portion of the video and oh no oh no OH HELL NO Lord Jehovah someone hold me like a baby, for I was scared. It was worse than I'd imagined. I yelped, hands flying over my mouth, watching a plane dropping through the sky looking as if it had lost power. Then the shot cut to two planes, noses pointed straight up in the air, in a side-by-side vertical climb. All the while, that voice in the background: *"Experience the rush of air-to-air combat!"*

Some of the video was shot inside the plane while other parts were outside as the planes filmed each other. These were the worst bits. There was an aerial shot of a plane blasting across the screen completely upside down. Another peeled off into a couple of barrel rolls, rotating wing over wing. One plane went into a backflip as an evasive measure but still got "shot down," fake smoke trailing out of the back.

But that wasn't all! In the final minute, we were treated to a sequence that ended with a plane in a nosedive hurtling toward the earth, spinning, spinning . . . Then the planes were landing. An instructor and customer shared a team-building high five. The narrator concluded: *"Give us a few hours and we'll bring out the fighter pilot in you! Do you think you have the right stuff?"*

I didn't. I had the wrong stuff. The *wrong stuff*! I returned the DVD to its sleeve, which had the company's motto printed across the top: *Everything is real . . . except the bullets!* This was not for people of reason.

"There's no way I'm going to be able to do this," I announced to Dr. Bob on the eve of my "mission."

"Have you ever noticed that some people love roller coasters while others are terrified of them?"

"Sure." I loved him for not saying, "Of course you can do this!" That's what everyone thinks they're supposed to say to show their support during moments of self-doubt. The problem is, you don't believe them. It comes off as a patronizing, naive platitude. On top of that, they are telling you that you are wrong, which I always find irritating.

"Your body can't tell the difference between fear and excitement. It reacts the same way to both—racing heartbeat, butterflies, perspiration. It's your mind that decides whether the situation is something to be nervous or excited about. What you need to do is turn fear into excitement."

Oh, is that all? "How?"

"Change your perspective of the situation. Change the narrative of your thoughts. Instead of thinking 'I'm so scared!' tell yourself, 'I'm so excited!' "

I eyed him dubiously. "And if that doesn't work?"

"Then try getting mad. One theory about anxiety is that it's opposed by other emotions, such as aggression, which can be used to cancel it out," he said. "Get into combat mode. Don't be a worrier, be a *warrior*! Imagine yourself as a flying predator going after the enemy. *Grrrr!*" He actually growled.

"I can barely handle being a *passenger* on a plane being piloted by a licensed professional. Even the slightest bit of turbulence freaks me out."

"What do you do during turbulence?"

"I grab on to the seat cushion and hold on tight." I was already gripping one of the arms of his sofa just thinking about turbulence.

"Which gives you the illusion of control in a situation where you've relinquished all control to the pilot. Doing it makes you feel safer."

"Isn't that a good thing?"

"When the plane doesn't crash, you believe on some level that grabbing the seat cushion is what saved you. Then you have to grab the seat cushion every time you feel turbulence."

"So I grab the seat during turbulence. What's the big deal?"

"It's what we call a safety behavior, like tensing up, holding your breath, praying, something we use to try to control the situation when we're scared, even if we can't control it. At best, these behaviors become superstitions. At worst, they can lead to substance abuse if you start believing that you need a few drinks or pills to get through a party. They disempower you by reinforcing the idea that you can't handle a situation."

"But I *actually can't* handle the situation that's going to be happening tomorrow! I don't know how to fly a plane!"

"Well, I'm no aviation expert, but you know what advice I would give to a first-time pilot?"

"What's that?"

"If you run into any turbulence, don't grab your seat cushion."

Later that night I watched *Top Gun,* bare feet propped up on the coffee table bouncing in rhythm to the cheesy awesomeness of Kenny Loggins's opening song, "Highway to the Danger Zone." The movie had aired a few weeks ago on TNT, and I'd recorded it specifically to watch tonight, for firing-up purposes. Instead I watched Goose eject into the canopy and break his neck, and I kept rewinding, watching him retake his seat in the plane, only to suffer his fate all over again.

The next morning Matt was driving me down the Long Island Expressway to Air Combat, which was based out of a small airport an hour and a half away. We were running late because I had done everything I could that morning to delay us, down to taking the full dentist-recommended three minutes to brush my teeth. I hadn't known it was possible to drive with a swagger until I'd met Matt. He eased his way around cars with the same confidence he used when striding across crowded rooms, knowing people would yield to him and automatically move to the side, which they did. Most of the time.

We'd just moved into the far left lane, only to be cut off by someone driving ten miles per hour slower than the speed limit. "Look at this douche bag," Matt said indignantly, thrusting a hand in the air at the driver in front of us as if slapping him upside the head. "I'm trying to get my girl out to her flying lesson and he's in the passing lane!"

I twisted my body in my seat toward Matt and suggested brightly, "Maybe we should go to Central Park instead! It's such a beautiful day, after all. Shall we turn the car around?"

"Sorry," he said. "I will not be party to any kind of wussery."

Scowling, I turned straight ahead and sank back into my seat in a huff. "Says the man who's afraid of heights! If I'm such a wuss, why don't you do it?"

"I would if it weren't so expensive. I'm not afraid of flying, I'm afraid of falling."

"Most fears are over in five minutes. I have to fly this plane for an hour and—wait, why are you laughing?"

"I'm sorry, it's just kind of funny. Most of your fears are about letting go of your inner control freak. Now you're doing something where you get to be in total control, but you don't like that, either."

"Listen, I wouldn't trust a first-time pilot no matter who it was," I argued. "For the same reason that I wouldn't trust a first-time brain surgeon to operate on me."

My phone rang. The caller ID screen said: *Mom*. She'd been driving me crazy about this, calling with more frequency as the date drew near. I picked up and before she could utter a word, I said, "Seriously, Mom, I cannot talk to you right now. I am trying very hard not to freak out and you are going to *freak me out*. I'm teetering on the edge here, Mom. Do you understand? *Teetering!*"

"You can still back out," she whimpered. I heard the muffled taps of her acrylic nails as she anxiously clutched at her phone. "You don't have to do this." Though a few minutes before I had been trying to talk myself out of this, her trying to talk me out of it somehow strengthened my resolve.

"I *do* have to do this! I made a promise. To myself. To Eleanor. To the universe. I'll call you when it's over." I hung up.

"That wasn't very nice," Matt observed.

"I know," I grunted. I'd regretted it even as I'd said it, yet I hadn't been able to stop myself. "But I can't deal. It's like getting a call from my own subconscious."

My mother was a worrier. After I'd gotten my driver's license, every time I'd left the house, my mom would say, "Watch out on the road. There are a lot of crazy drivers out there." Even now she rarely let me drive my little sister anywhere, afraid that if there was an accident she'd lose us both. *You can never be too careful* had been a common refrain when I was growing up. You could, actually. According to Dr. Bob, overprotective parents, in their attempt to raise conscientious children, were constantly sending the message that the world is full of dangers that will surely get you as soon as you let your guard down. Kids became trained to find risk in every new experience. My mom had drilled these warnings into my head when I was growing up; after I left home, her voice had become the voice in my head.

"Well, hanging up on your mom is pretty childish."

"Well, she treats me like a child!" I said petulantly. "How am I supposed to get past these fears if she's always calling and reinforcing them?"

But his comment only made me feel guiltier. Fantastic. Now I was worrying about her worrying about me. And somewhere she was probably worrying that she'd upset me. It was like a goddamn hall of mirrors up in here. I wasn't trying to push her away, but now that I was being trained to patrol my thoughts for worry, it made me aware of how often she did it. It was hard to keep the edge out of my voice whenever she started in. She was also extremely sensitive to criticism, and I'd gotten snippy with her enough times that she barely called anymore. In the past, we'd have long chats; then she'd pass the phone to my dad for a few wrap-up questions. Now it was my dad who called.

Matt changed the subject. "Aren't you the least bit excited, though?

You're about to do something you'll remember for the rest of your life."

"Oh, I'll remember this for the rest of my life," I said, "all three hours of it."

Matt gave up trying to rally my spirits and flipped on the radio. "Highway to the Danger Zone" blared forth.

"NOOOOO!" I cried in disbelief. For the first time that day, I was smiling.

"It can't be!" Matt exclaimed.

We fully rocked out. I was head-banging and playing air guitar. He was drumming on the steering wheel as he drove. The person in the next car over was probably saying, "Look at those douche bags!" But it didn't matter. We were in the zone.

An hour and a half later our shoes were crunching across the airport parking lot as we walked hand in hand toward the Air Combat office. Along one side of the lot was a chain-link fence and through its empty wire diamonds I could see a runway and a small airfield full of prop planes. We entered the unassuming one-story building that could've been a chiropractor's office in another life. An amiable receptionist at the front desk directed us to a room down the hallway. It was small and unadorned with two long tables and a few chairs. Standing at the front of the room in an army green flight suit was our instructor, a former U.S. Marine named Larry who wanted us to call him "Slick." This was his call sign, or pilot nickname.

"In addition to being an instructor, I'm also the company's mechanic, so I'm usually covered in oil," Slick explained. He shook our hands, and Matt asked where the restroom was and excused himself. The only other student in the room was a sweet-faced man wearing street clothes, sitting at one of the long tables where he was filling out paperwork and releases. The man looked up, his glasses reflecting the overhead light, and gave a little wave. "Hi, I'm Lenny." This was my "enemy."

"Have a seat," Slick offered. I sat at one of the tables. He smiled at me but said nothing.

Matt returned from the bathroom, and Slick handed him some papers. "Just fill these out, initial right here, and sign the bottom."

Classic. I smirked and cleared my throat. "Actually, I'm the one flying today."

He could barely conceal his surprise. "Oh! Okay, well, here you go." He handed the sheets over to me.

While I was signing my life away Slick asked Matt, "So did you buy this for her as a present?"

"No, she bought it for herself."

Another expression of bemusement.

"He's just here for moral support," I explained as Matt settled down with a book at the table behind me. Slick handed Lenny and me flight suits identical to his with American flag patches on the shoulder and long zippers up the middle spanning the neck to the nether regions.

As we headed to changing rooms to disrobe, he called after us, "It gets hot in the plane so you're going to want to wear only your underwear beneath these suits." Ground school had begun.

"Have either of you ever flown before?" Slick asked after we'd returned, suited up. I shook my head.

"I fly a lot of gliders," Lenny said, referring to the engineless planes that are towed up into the sky by another plane and then cut loose to sashay across the wind currents. "But they only go about sixty miles per hour."

"Today you're going to be going two hundred and thirty miles per hour," he said. "Here we fly Marchetti SF-260s, a plane used by several air forces around the world for fighter pilot training. These are the Ferraris of the sky. They're light and easy to maneuver."

He continued: "You and the instructor will be sitting side by side. Your control sticks operate together like the pedals in a driver's-ed car. When you're moving your stick, your instructor's stick will move as well, and vice versa. That way he can take over the controls at any time if you get into trouble." I glanced back at Matt, eyes wide. He gave a reassuring smile. When I turned back, Slick was holding up some kind of canvas sack with straps. It looked like a backpack from the 1800s.

"We've had over fifty thousand customers. We've never had a death, never had to use one of these parachutes," he said. "But just in case there's a fire in the cabin or a wing falls off . . ."

It can do that? Just FALL OFF?

". . . pull back the canopy and jump out of the plane. To open the chute, pull this D-ring right here." I took note of the parachute knowing that I'd never have to use it, because if anything happened that necessitated getting out of the plane, I'd have long since gone into cardiac arrest.

According to Slick, something else we wanted to watch out for was "buffeting." "Buffeting is when you feel the plane start to shake and you hear a loud continuous sound like *BAH-BAH-BAH-BAH-BAH*." He shook his fists in the air for emphasis. "It means that there's too little air pressure going underneath the wings and too much air pressure going over the top of the wings. It also means that the plane is about to stall." I moved my foot over next to Matt's, and he rubbed his shoe against mine.

Slick continued: "Now obviously when you're in combat, you do whatever it takes to kill the enemy, but for our purposes we have some rules. First, three thousand feet is the 'hard deck,' meaning that if you go below that altitude you immediately forfeit the fight."

"Hi there!" We were interrupted by a man in his early sixties sticking his head into the room.

"Lenny will be flying with me," Slick said and then nodded at the other man, who was also wearing a flight suit. "Noelle, this is your instructor, call sign 'Boom.' " Given my fear of crashing, I didn't have a great deal of confidence in an instructor named Boom.

After he ducked out, I raised my hand. "How do we keep from crashing into each other?"

"That's rule number two," Slick said. "Head-on approaches are not

allowed. You can only shoot your opponent from behind." He took out
two pencils, each with a tiny gray plastic plane perched on the end.

"F/A-18s!" Matt exclaimed, setting down his sci-fi book. "I used to
build models of those when I was a kid."

Slick nodded appreciatively. "This isn't what you'll be flying today,
obviously, but they'll do for demonstration purposes. Now the only
time you'll be facing each other is at the beginning of the dogfight." He
held up the planes, facing them toward each other, about ten inches
apart. "You'll fly toward each other, keeping your opponent on the left.
As soon as you pass each other, we'll say, 'Fight's on!' Then you try to
get behind the other plane and go in for the kill." Using the pencil
planes, Slick showed us some basic aerial maneuvers called "yo-yos"
and "lead and lag."

"At some point you might do this with the plane," Slick said, making one of the plastic planes do a backflip.

Uh, you *might do that with the plane,* I thought to myself. My ass
wasn't doing anything remotely similar to that.

He continued: "When you do, it's important that you go full throttle. Because if you half-ass the backflip or lose your nerve halfway
through, your aircraft will do this." The plane in his hand suddenly
dropped and started plummeting toward the ground nose-first.

This was the point where I decided I wouldn't be dogfighting.
My plan was to get a few hundred feet off the ground, freak out, and
demand to come down immediately. At the very most, I'd take over
the controls and fly the plane straight for a few minutes and then ask
Boom to take me back to the airport. A sense of peace came over me
now that I'd chosen not to do this. Still, I felt bad for Lenny, who'd
signed up expecting to have a dogfight with a fellow thrill-seeker.
Maybe after I was brought back down, Boom could go back up alone
and show him a good time.

"G-forces!" Slick boomed out. "G-force is the acceleration of an object relative to free fall. A negative g-force occurs when the plane is in a
dive. Your body feels lighter than it really is. A positive g-force occurs

when your plane is going up, multiplying the force of gravity and making your body feel heavier than it really is. Positive g-forces push blood away from your head toward your feet and can result in tunnel vision and loss of consciousness." In a two-g maneuver, my 125-pound body would feel like I weigh 250 pounds, at three gs I would feel like 375 pounds, and so on. At six gs, or 750 pounds, I would black out, which would probably be for the best.

The last thing he showed us was how to deploy the barf bag. Apparently, one out of ten customers vomited.

"The record is seven bags," Slick said proudly. "Guy went to a buffet lunch before he got here."

During the war, Eleanor lobbied strongly for starting a women's flying division in the Army Air Force. She argued that if more women took on domestic aviation jobs, more male pilots could be released for combat. "This is not a time when women should be patient," she wrote in 1942 in her newspaper column "My Day." "We are in a war and we need to fight it with all our ability and every weapon possible. Women pilots, in this particular case, are a weapon waiting to be used."

She also rallied behind African American airmen. In 1941, Eleanor visited the Tuskegee flying school in Alabama. Over the Secret Service's objections, the fifty-seven-year-old First Lady flew with a black pilot for more than an hour. The pilot, C. Alfred Anderson, later wrote in his memoir, "She told me, 'I always heard Negroes couldn't fly and I wondered if you'd mind taking me up.' . . . When we came back, she said, 'Well, you can fly all right.' I'm positive that when she went home, she said, 'Franklin, I flew with those boys down there, and you're going to have to do something about it.'"

There's a fantastic photo of the two of them in the two-seater plane. Eleanor is in the backseat wearing a hat with flowers on it, grinning broadly. The youthful Anderson is in the front looking pleased but nervous. "Please, God, don't let me kill this white lady," his expression is saying. But her plan worked. The symbolic value of the white First Lady sitting behind a black pilot was immeasurable. According to Anderson, the Army Air Corps began training African Americans several days after Eleanor's flight.

Boom and I were sitting side by side in adjacent airplane seats. The cockpit was small, like the front seat of a car, but with a clear plastic canopy over the top like the cars on *The Jetsons*. I'd just received a brief tutorial on how to fly. My instructions were few—don't touch anything but the control stick. It had a red button under my thumb, which I could press to talk to Boom on the headset attached to my helmet. Another red button let me talk on the radio to the other plane. There was a red trigger on the front, connected to the dashboard's computer. If you got the enemy plane (known as a bogey) in your gunsight and pulled the trigger, white smoke would come out of the tail, indicating a "kill." Otherwise, the control stick operated exactly as you'd imagine—push it forward and the plane went down, pull it back and the plane went up, move it right and the right side of the plane tilted toward the ground and vice versa on the left.

The engine started with a succession of *tdt-tdt-tdt-tdt*s and we were cruising down the runway. Charging. I was taking deep, calming breaths. We were up! I wasn't scared at all, which was odd because this was my least favorite part of commercial flights since I knew that 80 percent of crashes occurred shortly after takeoff. Somehow this liftoff felt natural. The plane didn't falter as it whirred over forest, sand, and then water, which blinked furiously under the unforgiving sunlight.

"So how long have you been flying?" I asked Boom.

"Spent twenty years in the navy. Flew attack missions in Vietnam. Then I went on to work for Pan-Am and a bunch of other places."

Did these military guys ever resent their guest pilots? I wondered. People who wanted the thrill of battle but would never enlist themselves. Then to have them walk away at the end, shaking their heads in a self-satisfied manner, saying, "That was fun, but I'd never do it for a living." Like the tourists who came to New York, fumbled around, asked us directions and if we'd mind taking their photo, and then left declaring, "New York is a great place to visit, but I wouldn't want to live there." (*What a coincidence!* I'd always thought. *I don't want you living here either!*)

"How did you get the call sign 'Boom'?"

"I never tell a story without a drink in my hand," he winked.

"I can respect that," I said. "And, might I add, I'm glad you don't have a drink in your hand."

He laughed. I liked this guy.

"So do I get a call sign?" I asked.

"Oh, you'll get one. At the end of the day."

The plane faltered a bit and we dropped a few feet and then reared back up violently. My stomach seized.

"That's just the prop wash," Boom said dismissively. Flying through another plane's wake, known as the wash, caused turbulence. And as I knew from last night's viewing of *Top Gun,* it was flying through Iceman's jet wash that had caused Maverick's flat spin and the demise of much-beloved Goose.

"Will we be expecting any more of those?" I asked nervously, eyeing the portable life preserver Boom had clipped to my waist before takeoff "just in case."

"Naw, probably not," he said, but he put some distance between our planes nonetheless. We'd been in the air for less than ten minutes when he said, "Okay, you have the plane." He let go of his control stick, palms facing out in one of those "look, Ma, no hands!" gestures.

"What?! Jesus!" The right wing started to drop, and I fumbled for my stick. I was filled with the same surprised indignation I'd felt when my dad taught me how to ride a bike. He'd been jogging along behind me, holding the back of my seat, and then without warning, he'd let go. "Stop it!" I'd shrieked. "Don't let go!" But there had been little I could do because it had taken all my concentration not to crash the bike.

"See? You're doing great!" Boom said as I struggled to even us out. The control stick was so sensitive that pushing it a millimeter was enough to move the aircraft.

When I was little, I liked to play on the seesaws at the neighborhood playground. Sometimes I'd stand in the middle of the seesaw, trying to balance both ends in the air at once. This was hard to do. I'd always end up putting too much weight on one foot, and when the seesaw started to drop on that side, I'd ease up and shift more weight to the other foot, and then *that* side would start to drop. This was what it was like trying to fly. I was trying to hold the control stick still, but the plane was somewhat cockeyed. So I nudged the stick slightly to the left. *Too much!* Back to the right again. The plane wobbled drunkenly. As I moved my control stick, I could see Boom's stick move with it as though guided by an invisible hand.

"Now we're a little high," Boom cautioned. "Push forward on the nose a bit." I urged the stick forward and the plane lurched. I overcompensated by pulling the nose up too fast. Eventually I straightened out. There was no way I was going to be able to dogfight. It was time to break the news to Boom.

"Um, I don't think I'll be able to, you know, do any tricks or go upside down or anything."

Boom waved this off. "Don't worry. When you're dogfighting, you'll be so focused you won't even notice when you're upside down."

"Trust me, I'll notice."

"Now I want you to get behind Lenny and practice getting him in your crosshairs. During the dogfight, he's going to be flying all over the place, so you won't look for him through your gunsight. Look for him out in the sky, maneuver the plane toward him, and then line him up in your crosshairs."

He added: "Always be looking for the bogey. Never lose sight of your enemy. You lose sight, you lose the fight."

I steered the plane over to Lenny's, and squinting one eye, I positioned him inside my orange circle with a cross in the middle. "Should I pull the trigger?" I asked.

"We're not activated yet, but why the hell not? It's emotionally satisfying."

I pulled the red trigger with my pointer finger, making *pew, pew, pew!* sounds. It *was* satisfying.

"Okay, now I want you to do a barrel roll. You're going to pull the control stick all the way to the left and keep going, flipping the plane over three hundred and sixty degrees."

Panic surged through me. "Oh, I don't think that's a good idea!" My voice was uncharacteristically girlish and fluttery. "Really, I'm fine right here. Can we just hang out?"

"To the left, now! Go! Go! Go!" he barked militarily, leaving me no choice.

With a stream of expletives running through my head, I pulled the stick over until it was pressed against my left leg and could go no farther. We were turning. The flickering gray ocean beneath the plane was replaced by the steady seamless blue of the sky. I felt a pronounced, but not uncomfortable, pressure on my body. Then the water rotated back into view and we were right side up again.

"Pretty cool, huh?" Boom grinned.

It . . . was, actually, I marveled. I couldn't believe I'd just done that. Now I understood how commanders got their soldiers to charge into battle. The human instinct to please could be more powerful than our survival instinct.

Boom had me hold the plane steady while we let Lenny practice getting me in his crosshairs.

"Time to dogfight!" Boom announced.

Oh no. There was no way out of this. Boom took the controls and steered me far away from Lenny and Slick's plane. When we were a good distance apart, he turned the plane around so Lenny and I were facing each other. We were two gunslingers moving toward each other on the main street of a deserted town. As we closed in, I took great care to stay toward the right side.

"Fight's on!" Slick's voice crackled over the radio.

Fwoom! The planes passed, left wing to left wing. Instead of turning and engaging with Lenny, I kept going straight. Maybe I could just outrun him? I *did* have a head start. *What this plane needs is a rearview mirror,* I thought, looking over my shoulder to see how far behind he was. It was bizarre taking my eyes "off the road." Then again, there was no one else up there that I could possibly crash into. Lenny was arching high across the sky, with obvious intentions to circle around behind me. Staring at the underbelly of his plane, wings jutting out like fins, I was reminded of scuba diving and how it felt to see a shark cruising overhead, knowing it was about to swoop down on me.

"Cut him off! Cut him off!" Boom shouted.

Cut him off? He wants me to cut off a plane? Using my plane?

"Bank left!" Boom ordered.

Cautiously, I eased the plane to the left.

"Harder! Harder!"

Clenching my teeth, I applied more pressure to the stick. The plane rolled violently sideways into a ninety-degree bank. The wings were completely vertical. I stole a glance out the left window. Eerily, all I saw was a wall of water.

"Tip the nose down!" he commanded.

I followed his order and suddenly we were dropping nose-first toward the ocean. My body felt empty. It was a horrible sensation.

"Oh, what's happening?" I asked in a rising tone of alarm. "Can you jump in here, please?!"

Boom was unfazed. "No, you've got it."

Instinctively, I pulled up on the nose and straightened out the plane, so that I was driving like a car, with the water beneath us. That was better.

"What are you doing? Don't look where you're going!" Boom repri-manded. "Look for the enemy!"

I blinked at him. The enemy?

"Lenny! Remember, lose sight, lose the fight!"

Right. Lenny. I braced my elbow on my headrest and shifted in my seat. I craned my neck, frantically scanning for a plane amid the blueness. Where the hell had he gone? You'd think he'd be easy to spot in an empty sky, but it was not like standing on a flat plain and look-ing for someone in the distance. He could've also been below me or above me.

Suddenly Lenny was upon us. While I had been struggling with the plane, he'd swung in behind and nailed me. White smoke poured out of my tail. The entire dogfight was over in a matter of minutes.

"Good kill," Boom congratulated Lenny over the radio. Then he turned to me. "You're doing a great job. Just remember, concentrate on watching the enemy, you can't hurt the plane."

"No offense, but I don't give a crap about the plane. It's me I'm wor-ried about hurting!"

He chuckled good-naturedly. "As long as the plane is safe, you're safe."

We lined up for the second dogfight. This time around I was less nervous as we charged toward each other. We passed, left wing to left wing; then Lenny fanned out to the right and I went left. For a second I was unsure of what to do next.

Boom hollered, "He's above us! Don't let him get away. Pull back! Pull back all the way! Pull! Pull!"

Slowly but firmly, I drew the stick toward me until it could go no farther. The earth dropped away as I steered the plane straight up. We were in a high climb, crusading against gravity. All I could see was the white blue sky and Lenny. Something was happening. What was this? I was pinned back in my chair, slumped over to the right, head tilted so far that it was practically resting on Boom's shoulder. I looked like someone should be wheeling me out during a telethon. The g-forces were so intense that I couldn't lift my head, not even an inch. I couldn't move anything.

"Keep pulling!" Boom was saying. I felt the g-forces coaxing the control stick out from under my fingers. If it popped out of my hand, we'd lose our thrust. All I could think of was the demonstration of what happened if you half-ass a backflip.

It was hard to speak, as if gravity was trying to keep the words down. In a feeble voice, I pleaded, "Help . . . me." That was all I could manage.

To my relief, Boom took over the stick. He steered us behind the bogey, lined it up in the crosshairs, and squeezed the trigger. White smoke streamed forlornly out of Lenny's tail. Boom hooted with pleasure. As he maneuvered the plane, I could only stare straight ahead. Clouds and ocean and sky twisted and turned in my eye line, as if through a kaleidoscope. Suddenly the pressure on my body ebbed. I could move again. I lifted my head and straightened in my seat and saw that we were flying right side up.

"Wow, that was intense," I breathed.

Boom gave me a happy nudge. "Congratulations. You won!" he said graciously.

My face was stuck in a huge, goofy smile. I didn't care that I hadn't pulled the trigger, I still felt like I'd won. *Holy shit,* I thought, *I'm actually doing this.*

There is no such thing as "down" during dogfights, I realized. The sooner you wrapped your head around that, the more success you'd have. Down was whichever way your butt happened to be pointing at any given time. You couldn't worry about where you were in relation to the ground. You had to keep your focus on the target.

"I got beat by a girl," Lenny moaned over the radio. I snickered even though Boom made the kill. Let him live in ignorance.

I was fired up as our third and final dogfight began. *Fwoom!* Lenny was a blur as we raced past each other.

"Do you see him?" Boom prompted in that teachery tone that implied he knew the answer but wanted me to figure it out for myself. I twisted around in the cockpit, skimming the sky for the bogey.

"Wait, where is he? . . . Oh, I see him!" He was off on the left below us, turning around so he could get behind me.

I rolled the plane to the side with the nose low and began an inverted dive. The ocean twinkled happily in front of me as I dropped toward it, but it was a nonentity. I was so focused that all fear was gone. Dr. Bob had been right. As long as I concentrated on killing the enemy, I stopped worrying about what my plane was doing. The bogey was the only thing that existed right then. I dropped until I was below Lenny; then I twisted the opposite way and pulled my nose up, climbing back up to his level. Now I was behind the bogey, fixing him in my crosshairs.

"You got it. Fire! Fire!" Boom said. I squeezed the trigger and got off a few rounds, none of which connected.

Suddenly Lenny drifted out of the gunsight. I unsquinted my right eye and look out the window. Now he was above us, over my left shoulder, and about to give chase.

My eyes narrowed. *Don't even try it, bitch.*

I did a high sweeping turn and pulled back on the control stick. My plane's nose tilted upward and we were in a climb. I zoomed over after the bogey, determined to make the kill this time. I was closing in. Almost got it . . . almost . . . There was a loud banging sound and the wings started shaking. Buffeting. I was too steep. The airplane was about to stall. Fuck. Fucking *fuck*. I quickly shoved the nose forward and we tipped forward suddenly and it felt as though we were on the crest of a roller coaster so I jerked the nose back up. When I got the plane level again, I looked around for Lenny.

"Okay, where did he go?" I asked.

"He's right behind us."

"Oh no!" I cried. "How do I get away from him?!"

"You don't," Boom said flatly. "He just shot us."

"Oh."

"I think it's about time to head home," Boom said, taking back the controls.

Already? That was it? But I was just getting the hang of it! I wanted to try another barrel roll. I didn't even care that I'd lost two of the three dogfights and that Boom had technically won mine for me. I had flown! Not only flown, but fought in air-to-air combat. And I hadn't freaked out or cried or asked to go home. I'd maintained possession of my stomach contents. I felt myself glowing. I could do anything. *Anything*. I was a warrior. Boom steered us in the direction of the flight school.

"Do you want to fly again?" he asked.

I nodded enthusiastically.

"Okay, you have the plane."

I was in the lead. I glanced over my right shoulder and saw Lenny following on my wing. He was less than ten feet away, but I was at full throttle so I couldn't speed up and put distance between us. *Why is he so close? Step off, Lenny! Oh, he's taking a picture of me. While he's driving. That can't be safe.* Still, I grinned and gamely gave him the thumbs-up, just so he'd turn his attention back to navigating. He took the photo, gave me a thumbs-up in return, and dropped back a little. When the airport bobbed into view, Boom took back the controls. We had to separate the planes so we could land.

"Say good-bye," Boom instructed me.

I pressed the radio button and said, "Byyyyeeeeeee—" but the word turned into a squeal when Boom banked sharply to the left.

"A little notice next time?" I asked.

He laughed and continued circling around toward the opposite end of the runway. We were maybe five hundred feet off the ground when the plane began to buck. Boom took a firmer hand with the control stick, wrestling with the beast.

"Thermal turbulence," Boom explained. "Caused by hot air rising off the earth. It's always worst in the afternoon when the ground is the hottest."

The plane was jerking, dipping. We were slaloming toward the landing strip. For the first time that day I was genuinely afraid for my safety. Instinctively I grabbed the bottom of my seat with my hands, but remembering my conversation with Dr. Bob, I immediately let go. An alarm in the cabin buzzed relentlessly: *Eh! Eh! Eh!*

Boom started flipping switches furiously. Oh shit. Something was wrong. After all that, only to be killed on the return? I cased the woods to our left. If we crashed into the treetops, maybe they'd soften our landing? No, it was going to be one of those fiery affairs, I could tell. Black smoke corkscrewing into the sky. Firefighters in silver suits. Local news helicopters muscling in for the best aerial shot to capture the carnage. Or maybe the force of the crash would eject me, Goose-like, from the plane? Only I would break through the Marchetti's flimsy canopy. "Are you sure there were two pilots?" the first officer on the scene would ask Slick. "There was only one body." They'd find me a week later lodged in one of the trees. Dental records would be procured for identification purposes, my face having already been eaten by wildlife.

Eh! Eh! Eh! Boom was still fumbling, not telling me what was going on.

Matt had been so cute earlier taking pictures of me—*the last photos!*—before we'd taken off. After I was gone, there would be news paper stories about me and the project, the sad irony of it all. It would scare the shit out of readers. Inadvertently, I would uninspire thousands. They'd take to their couches, eschewing bravery for television sitcoms and cop shows. I'd be the anti—Eleanor Roosevelt.

The noise stopped and Boom settled back in his chair. I exhaled with relief. In retrospect, the whole thing had lasted less than ten seconds.

"That was the automatic landing gear alarm," he explained. "If you go below a specific altitude and you've forgotten to lower the wheels, it lets you know." I could hear the wheels whirring into position. He chuckled. "Sometimes the plane is smarter than we are."

"I'm glad you didn't tell me that before we took off."

The wheels *eeeaaaak*ed onto the runway. As we were taxiing, Boom ripped back the canopy and, without realizing it, bonked me mightily on the head.

"Doesn't *that* feel good!" he exclaimed into the breeze.

As we rolled to a halt and climbed out of the cockpit I spotted Matt waiting off to the side on the runway. When he saw that I was grinning and not in need of sedation, he pulled out a video camera and peppered me with questions.

"Would you do it again?" he asked.

"I would, actually!" It was slightly embarrassing to admit. I'd spent weeks worrying and whining about this one hour. The scariest thing I'd done in my life so far turned out not to be all that scary—fun even. It made me wonder what else I was missing out on. Also, what other things in my life was I unnecessarily wasting time and energy worrying about?

"Was it scary?" Matt asked as I climbed down the wing.

"Nah," I said, then qualified: "Well, maybe a *little* during the landing . . ."

Matt took pictures of Boom and me, then of the two of us standing with Lenny and Slick in front of my plane.

"Can I get a few more shots of you two in front of the tail end?" Matt asked Boom. "Would you mind?"

"Take as much time as you want. We get all sorts of crazy requests," Boom said. "We've had women strip down to bikinis and pose lying on top of the plane. Someone else had us take a picture while she did a handstand on the wing."

Later, after we'd said our good-byes, I opened the car door to find a paper plate on my seat with half of a funnel cake on it. "I went to the car show next door," Matt said sheepishly. "Saved you some!" As he negotiated our way back onto the Long Island Expressway, I paused while licking the sugar off my fingers.

"Hey wait, I never got my call sign!"

"What?"

"Boom told me I'd get my fighter pilot nickname at the end of the day. He must have forgotten."

"So call him and ask."

I winced. I never liked calling strangers—a ridiculous admission for a former reporter, I know. Even as a kid, it had taken years before I could comfortably order a pizza. It got much better as I got older; then e-mail and texting arrived like manna from heaven for the telephone-challenged. In the last few years, especially, as my life shifted even more toward writing and the Internet, I'd regressed to being the child who wished I could ask a parent to call on my behalf.

I picked up the phone to call Boom several times over the next week. In the end, I chickened out. I went to the Air Combat USA website, e-mailed the webmaster and got Boom's e-mail address from her. After a few throat-clearing lines of "hey, remember me?" I finally asked the question: "So what's my call sign?"

A few days later I received a reply. "Despite your initial quavers, ya done good," Boom wrote. "Aggressive to the point where I had to throw a leash on you to keep from overextending. Hence, from henceforth, in fighter pilot circles, you shall be known as *Fearless*. Hope you do fly with me again, because afterward the cry will be, 'Fearless, you're buying the beer!'"

For a few moments, I just stared at the e-mail, smiling. I'd felt like a bit of a coward for e-mailing instead of calling, but now I was grateful that I could keep this conversation forever. Still, I wanted to atone for my sin. I picked up the phone, and when the person on the other line picked up, I said, "Mom? You won't believe the e-mail I just got . . ."

Chapter Seven

~

My life can be so arranged that I can live on whatever I have. If I cannot live as I have lived in the past, I shall live differently, and living differently does not mean living with less attention to the things that make life gracious and pleasant or with less enjoyment of things of the mind.

—ELEANOR ROOSEVELT

Fall continued apace. So did my Year of Fear with a random assortment of daunting tasks. I took a pole-dancing class. Jessica had refused to go with me ("I'm sorry, Noelle, but I have my limits"), but I left with considerably more respect for the hardworking ladies of the exotic dancing industry. Despite a lifelong fear of needles, I submitted to acupuncture, a horrifying one-hour event where I watched a man painfully insert needles into the tops of my feet and the tender webs of skin between my toes. I went back to trapeze school and spent two months training for a recital, where I performed—in costume—before hundreds of people.

I didn't wear makeup for two weeks. If this doesn't sound scary, you're not from Texas. In Texas, if you leave the house "without your face on," you might as well actually leave the house without your face on. People will react with the same level of revulsion. As a freelancer, I had days where the only person I saw was the guy who worked at the deli on the corner. Yet I found myself applying makeup to go order a sandwich. After more than fifteen years of wearing makeup every day, I realized, I'd come to think of my made-up face as my real face. Without makeup, I felt vulnerable, *less than*. So I decided to stop wearing it until I made peace with my face. My mom, who slept in her makeup for the first two years of my parents' marriage and to this day was always fully fragranced, was appalled. When she found out I went to a party barefaced, she said, "I wouldn't check the mail without makeup on, let alone go to a social gathering. That's just not—well, it's not *done*." It took two weeks, but I knew I was finally comfortable with myself when I spotted a former crush from college in a subway station and instead of ducking my head or flinching, I marched right up to him and, with rosacea and acne on full display, said, "Hey! Long time, no see!"

Now four months into the project, I decided that, to understand Eleanor, I was going to have to see her house. And to truly see Eleanor's house, I also had to see Franklin's house.

"Franklin's house?" Matt sounded confused as we climbed into bed. I was sleeping over at his apartment that night because, bless him, even though he'd driven in from Albany just a few hours before, the next day he'd be sacrificing his entire Saturday to accompany me upstate to visit the Roosevelt estate in Hyde Park, New York. "You mean they didn't live together?"

"They did for the first twenty years of their marriage. I'll explain on the drive up tomorrow."

Matt set his reading glasses on the bedside table and had just turned off the light to go to sleep when I suddenly threw back the covers and leaped out of bed. "Oh no, I haven't done my scary thing for today! I've been so focused on tomorrow that I forgot all about it." I pulled off my nightshirt—a black T-shirt Jessica had given me that said I SLEEP AROUND—and wriggled out of my panties.

Matt squinted at me through bleary eyes. "What are you doing? It's one in the morning!"

"I'm going to run down the hallway! Be right back." Running down the hallway naked had become my go-to if I got to the end of the day and nothing scary had come up. I'd yet to run into a neighbor, but it never failed to be scary. Before Matt could respond, I threw open his front door and tore out of his apartment.

The ride was two and a half hours, about as much time as I needed to explain to Matt the complicated living arrangements of this couple. Franklin's father was twenty-six years older than Franklin's mother, Sara. James Roosevelt was a widower with a grown son Sara's age. By the time Franklin came along in 1882, James was fifty-four and not much interested in parenting. He even had Franklin call him "Mr. James." But Sara tended to her son like a prizewinning orchid. She raised him at Springwood, the family estate on the banks of the Hudson River in Hyde Park. The house staff addressed the boy as "Master Franklin," and his smallest accomplishments were met with glowing praise. And while this kind of acclaim can produce disastrous results (children who overestimate their talents, enter the real world, and wilt at the first hint of rejection or criticism), it worked out brilliantly, not just for Franklin but for America as well. It takes a considerable amount of presumptuousness to believe you can lead a country out of a Great Depression. James Roosevelt died when Franklin was eighteen, causing Sara to cling even more possessively to her only child. She was cold toward Eleanor when the two announced their engagement, as she would have been toward any rival for her son's affections. When Eleanor moved in with them at Springwood, it was the stuff of bad sitcoms.

"Poor Eleanor," Matt said. "I wouldn't want to join that party."

There was no escaping her mother-in-law. As a wedding present, Sara built the couple a town house in Manhattan. However, this was one of those "gifts" where you give a family member a present you really want for yourself, knowing you'll have full access to it. For she'd also bought the plot of land next door and built an adjacent town house for herself that connected to theirs through sliding doors on several floors.

"You were never quite sure when she would appear night or day," Eleanor glumly recalled.

"It looks so . . . presidential," Matt said, gazing up at Springwood from the driveway. We had just strolled through a vast apple orchard and were standing in front of a colonial revival mansion with a tour group of forty Caucasians in casual sportswear. Our guide, a pink-complected woman named Meg, who was dressed like a park ranger, positioned herself on the stairs before the columned portico entrance.

"Franklin stood in this very spot after all four elections to greet the crowd that had come to congratulate him on his victory," she told us. "Since Hyde Park was a Republican district, he joked, 'I know you didn't vote for me, but I am glad to see you anyway.' " The crowd chuckled dutifully as we followed her into the shadowy foyer.

Franklin was a collector, according to Meg. His stamp collection totaled more than a million. He and his mother displayed his acquisitions proudly in the entrance hall. A group of framed political cartoons he found amusing hung neatly on the far wall. Nineteenth-century naval paintings adorned another. ("He served as assistant secretary of the navy for seven years," Meg reminded us.) Next to the door was a flock of stuffed birds Franklin shot as a boy. They were posed in flight to look like they were alive, which always struck me as missing the point; stranger still, his mother forbid the servants from touching them, tending to the dusting herself. If Franklin had died before her, I suspected she would've had him stuffed and then propped him up in a wingback chair. Instead she did the next best thing. Seated in front of the birds was a life-size bronze statue of Franklin that Sara commissioned when he was elected to the state senate at age twenty-nine.

Matt and I shuffled down the hallway behind the other tourists. The rooms were cordoned off with small gates, but you could peer into them as if it were a life-size dollhouse. We waited our turn to read the plaques explaining what room we were looking at and its significance. On our left was a cave of a den Sara called her "snuggery" where she conducted the business of the house. The furniture was too large for the space and, like Sara herself, had a way of making people feel small. One person, of course, took the brunt.

"Your mother only bore you," Sara once told Eleanor's son Jimmy. "I am more your mother than your mother is."

Add to that the occasional public insult, such as announcing to Eleanor during a dinner party, "If you'd just run your comb through your hair, dear, you'd look so much nicer."

Everything you need to know about the Roosevelt family dynamic can be found in the seating arrangements. Sara and Franklin sat at the heads of the dinner table. In the imposing wood-paneled library, a pair of matching, handsomely upholstered chairs sat on either side of the fireplace. One was for Franklin, the other for Sara. God knows where Eleanor sat, probably on the banks of the Hudson River trying not to throw herself in.

"For almost forty years, I was only a visitor there," she later wrote of Springwood.

In 1918, Franklin returned from a trip stricken with pneumonia, so Eleanor unpacked his bags for him. Inside she found a bundle of love letters addressed to Franklin. She recognized the handwriting instantly. They were from a woman named Lucy Mercer, Eleanor's secretary. It was yet another example of Franklin's audaciousness. It's one thing to cheat on your wife with your secretary, but someone who cheats on his wife with his *wife's* secretary is clearly operating on a different level. At that point he and Eleanor were thirteen years into their marriage; she'd borne him six children (one died as an infant). Eleanor offered to grant Franklin a divorce, but Sara intervened, knowing that such a scandal would ruin her son's political career. She threatened to disinherit Franklin if he went through with it. The Roosevelts decided to stay together on two conditions set forth by Eleanor: Franklin had to break off his relationship with Lucy Mercer immediately, and he could never share his wife's bed again. Ironically, only after that did Eleanor feel secure enough in her marriage to finally assert herself with Sara. She started, awesomely, by blocking the sliding doors connecting the twin town houses with all the heavy furniture she could find.

The Roosevelts spent the summer of 1921 at their summer home on Campobello Island off the coast of northern Maine. One afternoon, Franklin returned from a swim complaining of chills and back pain and went to bed early. "By the next morning, he could hardly stand, and by the next day, he could not stand at all," Eleanor remembered. She slept on a couch in his room and nursed him for nearly three weeks, but nothing could be done. Franklin had contracted polio and was paralyzed from the waist down.

Back at Springwood, portable ramps were placed throughout the house and pulled up before guests arrived. Franklin hid his paralysis in public to avoid seeming weak, sometimes even wearing leg braces that locked at the knee, which allowed him to stand upright. He gave the illusion he could walk by using a cane or leaning on someone's arm while using his hips to swing one leg forward at a time. Even during his presidency, he kept up appearances. When leaders came for meetings, he was already settled in his chair and stayed seated until they left. The media knew he was confined to a wheelchair, of course, but didn't feel it was appropriate to "out" the president on a personal matter. They only photographed him when he was sitting in a car, behind a desk, or leaning against a railing while delivering a speech.

He thoroughly charmed the press corps with his quick wit and naughty boy personality. At cocktail hour he corralled reporters and staff members in the cloakroom beneath the stairs because Sara disapproved of drinking in her house.

"It was made into fun," one journalist later recalled. "With shrieks of laughter we'd gather with the President of the United States, the coats hanging up on the wall, he in his wheelchair whipping up the martinis and drinking as if we were all bad children having a feast in the dormitory at night."

Having fully explored the first floor, our group drifted back to the main staircase where we took turns peeking in Franklin's hand-powered elevator, originally installed so the servants could haul the Roosevelts' heavy trunks upstairs after trips overseas. Franklin disliked standard wheelchairs so he designed his own, a regular wooden chair with wheels at the base of the legs. One of these sat inside the elevator taking up most of the compartment.

"Good thing he wasn't claustrophobic!" a dad guffawed loudly from beneath his baseball cap.

"It operates on a rope pulley system," Meg explained. "Franklin used his arm strength to raise and lower himself." As I looked inside, I was reminded of one of Franklin's most famous quotes: "When you get to the end of your rope, tie a knot and hang on."

"Couldn't he afford a motorized one?" the dad asked.

"FDR refused to have an electric elevator installed. He feared that in the event of a fire, the power would be shut off and he'd be trapped and burn alive. Fire was the one thing he feared because he couldn't run from it."

Meg divided the group so we could tramp upstairs in shifts to the second floor. Matt and I maneuvered ourselves in front of a family with six young children. This mansion should have been airy with its glossy white walls and soaring ceilings, but the navy carpet sucked in the light. It was stifling in spite of its thirty-five rooms and nine baths. We were all a little creeped out by the Birth Room, which still housed the bed where Franklin was delivered. Sara's deathbed request was that the room be rearranged to look exactly as it had when he was born in 1882. We hurried on to his sizable suite at the end of the hall. The bedroom was left intact after his final visit two weeks before he died. The customized phone with a direct line to the White House waited expectantly on the bedside table. Franklin's books and magazines remained strewn about the room where he'd left them. It was an eerie sight, like viewing the scene of a crime. I tried to imagine what my room would look like if I'd died suddenly and Meg started giving tours through my apartment. ("As you can see, her Chia Pet and the ass print on her sofa cushion are exactly as she left them . . .")

There was very little to see when I stuck my head into Eleanor's minuscule quarters next door, a converted dressing room. It was startling in its bareness, just a daybed, no private bathroom to call its own.

The third floor was closed to the public, so Matt and I ventured out onto the veranda and admired the view. The house sits on a hill, looking haughtily over the estate's six hundred acres. We lumbered down the outside staircase and back to the front of the house where Meg was waiting. Eleanor once noted that Franklin's polio was "a blessing in disguise, for it gave him strength and courage he had not had before. He had to think out the fundamentals of living and learn the greatest lesson of all—infinite patience and never-ending persistence."

As if to illustrate the point, before seeing everyone off, Meg pointed down the lengthy driveway. "If you want to know what it takes to be president, consider this," she said. "FDR came out here every day on his crutches and tried to make his way a quarter mile down the driveway to the main road. He felt that if he could make it to the end unassisted, one day he'd be able to walk again. If he fell, he'd lie facedown in the road until someone happened upon him and helped him back up. He never made it the whole way, but he still tried."

Most of the group headed back toward the presidential library, but Matt and I wandered through the orchard, enjoying the late fall sun.

After a few minutes, Matt stopped and looked around. "So which way to Eleanor's house?"

"We should probably drive, considering she lives two and a half miles away."

"Two and a half miles?" he repeated as we headed toward the parking lot. "Not exactly subtle."

When Franklin was stricken with polio, he'd been in politics for ten years as a state senator and the assistant secretary of the navy and had a failed bid at the White House as the Democratic vice presidential candidate in the 1920 election. He refused to accept that he was paralyzed and spent much of the 1920s doing physical therapy. To keep Franklin active in politics, Eleanor had to be his legs and his voice. She became active in organizations, made speeches. While volunteering for the women's division of the Democratic Party, she became best friends with longtime lovers and political activists Marion Dickerman and Nancy Cook. Franklin liked the couple too. One day, while the foursome picnicked by Fall Kill Creek, he proposed building a cottage on-site where the women could live full time.

"That's a little"—Matt paused before easing the Buick onto the main road—"unorthodox, isn't it?"

"He knew she needed to escape from Sara. I'll bet he was sick of the tension in the house himself."

He hired an architect, and by 1925 the women had a one-and-a-half-story fieldstone house overlooking the creek. Franklin even put in a dam to redirect the water into a swimming pool. Eleanor referred to Stone Cottage as their "love nest." The women traded everything from bathing suits to lipstick. They monogrammed their initials—EMN—on the linens and towels.

"I can honestly say I've never looked at any of my guy friends and thought, 'Let's make this official in terry cloth,'" Matt joked.

I told him how the women built a second structure on the property, Val-Kill Cottage, and started a furniture factory so unemployed locals could earn a supplemental income. Nan supervised the business while Eleanor taught literature, drama, and American history at a school in Manhattan where Marion was vice principal.

"And do those who monogram together stay together?" Matt asked.

"Not exactly. The factory went under in '37. Then one night when Marion was in Europe, Eleanor and Nancy had a terrible argument. Eleanor immediately moved out of Stone Cottage, and Marion and Nan later moved to Connecticut."

"What was the fight about?"

"No one knows. Though Marion later said it involved 'things that should not have been said.'"

"So?" Matt prompted. "Was she or wasn't she?"

"I honestly don't know. She did have a number of close lesbian friends . . ." People had been gossiping about this question since the 1920s and 1930s. Eleanor's own cousin, the mischievous Alice Roosevelt, once commented loudly in a chic Washington restaurant, "I don't care *what* you say, I simply cannot *believe* Eleanor Roosevelt is a lesbian."

I told Matt that in 1978, the staff at the Franklin D. Roosevelt Library opened eighteen boxes containing sixteen thousand pages of letters between Eleanor and Lorena Hickok, a thirty-five-year-old gay Associated Press reporter who smoked cigars and wore flannel shirts and trousers that brought to mind a lumberjack. Hick was assigned to cover Eleanor but became her confidante. She gave Eleanor a sapphire ring, which Eleanor wore at Franklin's 1933 inauguration. Eleanor sent her a note after the ceremony that read, "Hick, darling, I want to put my arms around you. I ache to hold you close. Your ring is a great comfort. I look at it and I think she does love me, or I wouldn't be wearing it."

Hick wrote to Eleanor, "I remember your eyes, with a kind of teasing smile in them, and the feeling of that soft spot just northeast of the corner of your mouth against my lips."

"Okay, hold up," Matt said. "Is there really a question here?"

I shrugged. "Not necessarily. Historians have pointed out that Victorian women wrote love letters to platonic friends because they were so starved for romance. Besides, some of Eleanor and Hick's letters suggest the feeling wasn't mutual."

In 1937, Eleanor wrote to Hick, "I know you have a feeling for me which for one reason or another I may not return in kind, but I love you just the same."

Matt followed a sign for the Eleanor Roosevelt National Historic Site and pulled the car into the parking lot.

"My feeling is, who gives a shit if she was a lesbian?" I said, closing the car door behind me. "Why does that matter? The woman ditched her harping live-in mother-in-law, built her dream house, and brought along her favorite gays. Personally, I think that's kind of awesome." Eleanor was ahead of her time. Back in 1925, she wrote in her diary, "No form of love is to be despised."

From the parking lot, we followed the curved path through the trees and came to a brook with a wooden plank bridge. Below, Fall Kill Creek murmured over the rocks. On the other side there was a pond, glass-still and lined with bushy trees that left their reflection imprinted on the water.

This time our guide was a librarian type in regular, non-ranger clothing.

"A lot of people ask where Val-Kill got its name. Don't worry, we're not a bunch of murderers up here!" She tittered, and I suspected she'd said this line hundreds of times. "This area was settled by the Dutch and 'kill' means 'little stream.'"

If Springwood was a twenty-thousand-square-foot thirty-five-bedroom fortress, Val-Kill Cottage was a bungalow. Its seven rooms were tacked on over the years, giving it a hodgepodge feel, as if an architect let all five of his multiple personalities have a go. Shaped like a bent arm, the front door was tucked into the crook.

"Visitors always thought that the front door was the back door," the tour guide noted as our group of ten traversed the threshold. The interior was encased in wood paneling, but a few screen porches kept it from feeling stuffy.

"When the weather was agreeable, Eleanor bunked on a sleeping porch upstairs," the guide informs, "where she liked having 'only the stars to look at, just because it gives one a feeling of taking in.'" I tried to imagine her lying there on this bed, waking up in the fresh air (she woke up at eight A.M. every day, no matter what time she'd gone to sleep). But whenever I toured historical homes, I could never picture their famous owners puttering around in them.

The rooms were, like Eleanor, welcoming and informal. The First Lady's official bedroom was a twin bed with a simple chenille bedspread, but with more photos and personal items than its Springwood counterpart. Springwood had been released to the government after Franklin's death in 1945 and Eleanor had returned to Val-Kill. For almost twenty years, dignitaries still called on her and sat around on the cushy chintz chairs in the living room to knock around ideas about affairs of state.

"Eleanor entertained the likes of Winston Churchill and Gandhi here," the guide said as we crowded in front of the roped-off casual dining room. "She often served them herself from this here side table." Winston also swam in the backyard pool while smoking a cigar, which made me like him even more.

Unlike Franklin who used his house to showcase his hobbies, Eleanor covered her walls with photos of friends and family, the people she collected over the course of her life who were important to her. On her walls, grandchildren, her secretary, and secretaries of state all mingled together. Her house had no trappings of power. The TV room, with its 1950s television and slipcovered armchairs, also doubled as her office, her desk occupying only a small corner. The nameplate read ELANOR ROOSEVELT.

"Why is Eleanor's name spelled wrong?" a familiar voice piped up. The dad from Franklin's house.

"A little boy made that for her in his woodshop class and she didn't have the heart to tell him he'd misspelled it. When guests asked why she kept it displayed so prominently, she replied, 'In case he comes back to visit one day.'"

She never stopped writing. She drafted most of the Universal Declaration of Human Rights here after Truman appointed her to the United Nations. When she died of bone cancer at age seventy-eight, she was in the middle of writing a book, *Tomorrow Is Now*. Altogether, she churned out seventy-three hundred newspaper columns and twenty-seven books. I couldn't fathom that kind of output. If I sat down at my computer when I got home and didn't get up for fifty years, I don't think I could produce that much writing.

Because the house was small, the tour went by much more quickly than the one at Springwood. As Matt and I retraced our steps down the driveway, I slipped my arm through his. Of course, we knew about living in separate houses. When Matt's editors had deployed him to Albany two years before to cover the state government beat, our relationship entered a holding pattern. Because he didn't live in Manhattan, we never had to deal with the issue of moving in together. Or getting married. Everything remained exactly the same, frozen in time like Eleanor's house. But how long could we keep this up? I wondered. How long until we became emotionally distant as well? What if he moved back and we realized we preferred our part-time relationship?

In the end, Eleanor came back to Franklin. She was buried next to him and their dog Fala in Sara's rose garden at Springwood. The couple shared a gravestone, a modest rectangular block bearing both their names. FDR always said he didn't want a gravestone bigger than his desk in the oval office. Walled in by tall hemlock hedges, Franklin and Eleanor could finally be alone. Sara, who died only three and a half years before her son, was buried off premises at a local church. That had to be something of an indignity, I imagined, being outranked by a Scottie. There's a Sara Delano Roosevelt Memorial Park in New York, which I had happened upon during a recent walk around my neighborhood. It was just a few blocks from my apartment. Eleanor was right: You never knew when she was going to appear.

Franklin and Eleanor's children went on to lead fairly turbulent lives. Elliott, bizarrely, wrote a mystery series starring his mother as a detective. He angered family members by penning a trilogy of revealing books about his parents, including details about their sexual lives and FDR's affairs. Between the five of them, the siblings had nineteen marriages, fifteen divorces, and twenty-nine children. Anna married three times. Franklin Jr. and Elliott each had five wives. James fathered seven children with four different wives and made headlines when his third wife stabbed him during a domestic argument. John was only married once but—perhaps most disturbingly of all for the Roosevelts—became a Republican. Yet Eleanor's support never wavered.

"No one ever lives up to the best in themselves all the time," she said, "and nearly all of us love people because of their weaknesses rather than because of their strengths."

As we walked back to the parking lot, we paused under the shade of the trees on the bridge over the dam Franklin had built for Eleanor. The creek happily burbled away on our right, teeming with energy, the lily pond earnest and steady on the left.

Matt leaned against the railing. "How much they did—the sheer productivity—just blows me away," he said, shaking his head. "I really felt that I was transported back to this period where greatness wasn't what you owned, it was what you *did*."

Standing there, staring into the water, I knew I needed to recommit myself to the project. The Roosevelts dedicated themselves to public service, and I was running naked through apartment buildings and taking stripper classes. Obviously, I needed to get more serious with my challenges. And maybe find a way to get more outside of myself as I do them. I'd never been in a position where I was offering something truly useful or important to others.

"To be useful is, in a way, to justify one's own existence," Eleanor said.

When I got home, I did an Internet search of the words *New York* and *volunteer*. The first website that popped up was a local hospital looking for volunteers. Perfect. Eleanor began volunteering as a child, accompanying her aunt Gracie to visit handicapped kids at a hospital in Manhattan. At the end of World War I, she visited the naval hospital in

Washington, D.C., once a week, bringing flowers and chocolates, and words of cheer. This contact with wounded soldiers, Eleanor later said, taught her an important lesson: "I was beginning to feel pity for the human condition. I was beginning to ask what I could do."

I downloaded the application and filled it out. The essay portion asked: "Why do you want to be a hospital volunteer?" I quickly banged out two hundred words: "My thirtieth birthday is approaching and as I look back at my life so far, I am ashamed at what I see. What stands out are not the things that I've done but the things I have failed to do. How can it be that I have been on this earth for almost thirty years but done nothing to help other people? When I look back at my life, I want to see a person who helped make other people's lives better . . ." I printed the essay out, read it over. Was it too over the top? Before I could second-guess myself, I stuffed the essay and application into an envelope, threw on my coat, and headed out to the mailbox at the end of my block.

Chapter Eight

～

The greatest thing I have learned is how good it is
to come home again.

—ELEANOR ROOSEVELT

When I arrived in Texas for Christmas, my Year of Fear was
almost halfway over. My first night home, my dad invited
the Valby family out to dinner at an Italian restaurant
renowned for its pagers that lit up like spaceships to signal diners
their table was ready. Mr. Valby was my dad's tennis partner. He was a
cheerful man in his late fifties who loved to hear about my life in New
York. When I told him about the project over dinner, he was fascinated.
As Mr. Valby asked question after question, I noticed my dad's head
going back and forth between us, looking increasingly displeased as if
he were witnessing the world's worst tennis match.

Finally Dad cleared his throat and said, "Her mother and I think
it'd be better if Noelle moved back home for a while, get her head on
straight. Maybe help out at my office."

Oh God. Could it come to that? What if my Year of Fear bankrupted
me and I had to move in with my parents in Sugar Land, Texas? The
shame! Oh, how I'd teased Mom and Dad when I was in college and
they announced they were moving from Houston to Sugar Land! "Is

that next to Candy Land?" I'd asked. "How far is your house from Gum Drop Mountain? Will you have to drive or can you just take the rainbow trail to get there?"

My parents are actually lovely people. They're supportive but incredibly practical. Having come from a long line of hardy farmers and businessmen of modest means, they had a hard time grasping that a person could make a living as a writer. For years they'd tried to steer me toward law school or a career in dentistry.

"You'd make enough money to support yourself," my mom would say, sounding slightly dreamy. She'd been a devoted stay-at-home mom, but I'd always sensed she regretted not forging a career for herself. "A woman shouldn't have to depend on a man for her livelihood," she'd often told me.

After I graduated from college, they'd watched nervously as I'd toiled away at newspapers on a $25,000-a-year salary while living in a very expensive city. They'd been thrilled for me when I'd landed the high-paying blogging job. Telling my parents I'd been laid off was harder than *being* laid off. As I'd picked up the phone to call them with the news, they'd felt more like my children than my parents. I'd wanted to shield them from the disappointment I was about to cause.

"Well," Mr. Valby's jovial drawl cut through the tension, "I still think this project of yours sounds mighty interestin'. You know what you should do?" He speared a shiny chunk of steak with his fork.

"What?" I asked.

"You should climb Mount Kilimanjaro. Fourth-highest mountain in the world, you know," he added, before depositing the steak in his mouth.

"Oh, that sounds dangerous. I don't think it's a good idea," my mother said, though I was pretty sure that, like me, she knew almost nothing about Kilimanjaro.

"Doesn't that kind of thing take years of preparation?" I asked. With only six months left on the project, I didn't have that kind of time. "I

don't have any mountain-climbing experience." I was picturing looming cliffs of ice, complicated rope systems, and frozen appendages that
turned black and had to be self-amputated with a Swiss Army knife.

"You don't need climbing experience for Kilimanjaro because it's
not really a climb," he said. "The mountain is so broad you're essentially just walking to the summit in a couple of days. Kilimanjaro requires no technical skill."

"That's me," I deadpanned. "Technically no skill."

Mountain climbing combined my two least favorite things on this
planet—camping and exercise. Throw in crapping in the woods and
that's a pretty accurate description of how I imagine hell. Still, I wondered if the universe was telling me—in the guise of a middle-aged
tennis enthusiast—that I needed to go mountain climbing.

When I got home from dinner, I plopped in front of my parents'
computer to find out more about Kilimanjaro, like, for instance, what
continent it was on. Answer: Africa. The closest I'd been to visiting
Africa was on the It's a Small World ride at Disneyland.

Mr. Valby was right. Climbing Kilimanjaro didn't necessitate prior
mountaineering experience. But the level of difficulty was a matter of
much debate. Reading hikers' testimonies about climbing Kilimanjaro was like asking a group of Democrats and Republicans what they
thought of the president of the United States. Half the hikers shrugged
off Kilimanjaro as a long walk. A disturbing number of people had
climbed it on their honeymoon. Others claimed it was the hardest
thing they'd ever done, mentally and physically. More than twenty-five
thousand people tried to climb Mount Kilimanjaro each year, but only
40 percent made it to the top. Fifteen thousand turned back before
they reached the summit. These numbers gave me pause. That was a
lot of people getting their asses kicked by a mountain. Was I just setting myself up for failure? Or something even worse?

Kilimanjaro offered a diverse and riveting selection of ways to
die: malaria, typhoid fever, yellow fever, hepatitis, meningitis, polio,

tetanus, and cholera. Those, of course, could be vaccinated against. There was no injection to protect you from the fog, which could roll in fast and as dense as clouds. According to one hiker's online testimonial, "At lunch . . . the fog was so thick, I did not know what I was eating until it was in my mouth. Even then, it was a guess." With zero visibility, people wandered off the trail and died of exposure. Even on a clear day, one could step on a loose rock and slide to an exhilarating demise. Or sometimes the mountain just came to you. In June 2006, three American climbers had been killed by a rockslide traveling 125 miles per second. Some of the boulders had been the size of cars, and scientists suspected the ice that held them in place had melted due to global warming. On the other end, hypothermia was also a concern. Temperatures could drop below zero at night. Then there was this heartening tidbit I came across in my research:

"At 20,000 feet, Mount Kilimanjaro is Africa's highest peak and also the world's tallest volcano. And although classified as dormant, Kilimanjaro has begun to stir, and evidence suggests that a massive landslide could rip open the side of the mountain causing a cataclysmic flow of hot gases and rock, similar to Mount St. Helens."

A volcano?! They're still making volcanoes?

But the biggest threat on Kilimanjaro was altitude sickness. It happened when you ascended too quickly. Symptoms could be as mild as nausea, shortness of breath, and a headache. At its worst it resulted in pulmonary edema, where your lungs filled up with fluid (essentially, drowning on land), or cerebral edema, where your brain swelled. Eighty percent of Kilimanjaro hikers got altitude sickness. Ten percent of those cases became life threatening or caused brain damage. Ten percent of 80 percent? I didn't like those odds. Maybe this trip *was* too dangerous. My eyes were stinging from staring at the computer too long, so I shut it down.

I wandered into my little sister's room, knowing she was away at a swim meet, to look for clues about what she'd like for Christmas. The walls had been painted bright blue since my last visit; they were full of photographs of her with friends I didn't recognize. Jordan had been a late arrival to the family, born fifteen years after me. I was in the delivery room when she arrived and even cut the umbilical cord (my father, a squeamish man, enjoyed the proceedings behind the safety of a curtain). I'd assumed we'd be best friends because we were too far apart in age for sibling rivalry. I'd been too young to understand that this age gap essentially guaranteed we wouldn't be close. She'd been a toddler when I'd left for college; since then she'd only seen me a few times a year during holidays.

On one wall there was an expansive cork bulletin board with her swimming medals dangling from ribbon necklaces on hooks along the bottom. They hung at least seven feet across. She was only fourteen but already ranked in the top ten nationally. Looking at that bulletin board, I felt very proud, of course, but I ached for her as well. There was something a little heartbreaking about this curtain of medals. At this point, it was more surprising when Jordan *didn't* win first place. I wondered if she'd gotten to the point where the fear of losing far outweighed the joy of winning.

"Oh, you startled me!" my mom said from the doorway where she stood, hand over heart, and smelling of sunflower-based perfume. "I didn't know you were in here. I'm just looking for the scissors. Jordan always takes them and forgets to put them back." She bustled into the room, shaking her head in exasperation. Sure enough, when we opened a desk drawer there was a pile of scissors.

"Oh, for Lord's sake!" Mom exclaimed, but her tone was affectionate.

Mom ran her fingers anxiously over the gold choker she always wore. "Do you think she's okay at the swim meet?"

"Of course. Mom, this is, like, her millionth swim meet away from home. The coaches are with her."

"I think I'll call again just to make sure. Anyway, I need to remind her not to stay up too late and to eat lightly at breakfast because she could get a cramp during the race and—"

"Okay, you've gotta stop doing that!" I burst out.

"Doing what?" Mom looked confused.

"Worrying all the time."

"I'm just trying to protect her." I didn't want to have this conversation. My mom's entire life was about being a mom, and I was about to tell her she had been doing part of it wrong.

"You think you're helping her, but you're really just making it harder for her in the long run. She'll grow up not knowing how to handle setbacks and disappointment. It's good for her to fail once in a while," I said, talking about myself as much as my sister. "Failure is a better teacher than success."

I took a deep breath and continued: "You have to give her freedom to fail once in a while so she can learn to give *herself* freedom to fail."

My mom's eyes filled with tears, but I had to finish this.

"And I'm not talking just about her, Mom. When you're always expressing doubt about the things that I'm doing, it makes me feel like you don't have confidence in my abilities. Which makes me doubt myself more."

That did it. Tears spilled down her cheeks as she said, "I do it because I care!"

At the sight of her tears, the frustration I'd been feeling toward her in the last six months dissipated. Worry was how my mother loved, I realized.

"But you're teaching her to equate caring with worrying," I said gently. "She'll grow up thinking that if you really care about your career, you should always be worrying about work. Or that if you really care about your relationship, you should worry whether your partner is cheating on you or falling out of love with you. That's not an easy way to go through life. I know you don't want that for her."

"No, you're right. I need to trust you girls more." She looked up at the ceiling while she wiped the tears away, to avoid smearing her mascara. "I just—I just like having people to take care of, you know?"

"You could always have another baby. Jordan's almost the age I was when you had her."

This got a laugh. "Oh, right! Could you *picture* the look on your father's face?"

"No, because I've never had someone stroke out right in front of me before."

She started laughing uncontrollably, and soon I joined her. Finally, we'd mostly settled down, with just a few aftershock giggles.

"It wasn't even that funny," I said. For some reason, this made us explode all over again.

The next morning as I was checking my e-mail, the phone rang. When I picked it up, Jessica started in without saying hello, a gesture that should have been rude but always gave me a sort of heartwarming thrill.

"So apparently," Jessica reported, "when I was stumbling home from our office holiday party last night, I bought a Christmas tree."

"Why do you say 'apparently'?" I asked.

"Because I'm staring at a fully decorated tree in the middle of my living room with no recollection of how it got there."

"You *decorated* while under the influence?" I asked. "Impressive. Did you also convert from Judaism to Christianity?" Jessica was Jewish.

"As drunk as I was, anything's possible."

I propped my feet on the edge of the desk and leaned back in my chair happily. It was one of those spinning chairs you'd find in an office.

"And what's the latest from Texas?" Jessica asked.

"I'm reading about Mount Kilimanjaro, actually. A friend of my dad's suggested it for my Year of Fear, and I thought it could be a good idea until I learned about the many ways one can meet his or her un-timely death while mountain climbing."

"Kilimanjaro? That's a great idea," Jessica said, with much more enthusiasm than I'd expected. She'd completely ignored the part about the potential for dying. "It's supposed to be a life-changing experience. Did you know you can see the curve of the earth up there? I've heard the sunrise is like nothing you've ever seen!"

"Seriously?" I'd been absently twisting my chair back and forth, but this made me stop and sit up. "I've never pictured you as the roughing-it kind."

"I know, but lately I just feel so stuck in my life in New York that I want to do something that's completely the opposite experience of living here. I'd do it myself, actually, if it weren't so expensive."

Uh-oh. Already I was more than halfway through my savings. Money was becoming an increasingly important factor in what I chose to do. "How much does it cost?"

"Thousands."

My heart sank. "Is that with or without airfare and hiking equipment?"

"Without."

"Well, there goes that idea." I pushed off the desk with one foot to make the chair spin around a few times. "No way I'll be able to afford that and have enough money to finish the project." As I twirled I saw something large blur by. I put my feet on the floor. "Oh! Hi, Dad. Jess, let me call you back."

He was wearing his long monogrammed flannel robe and slippers. My dad would never walk around the house in boxer shorts or with pa-jama pants and an undershirt. Everything in his manner—from the way he ate, spoke, and dressed—carried a certain dignity. In fact, I couldn't recall him ever wearing a shirt without a collar, or blue jeans. When he was in his robe and slippers, it was the only time he appeared vulnerable to me.

"I didn't mean to interrupt. I just thought I'd ask if you wanted to go shopping for your mother's and sister's Christmas presents together tonight." He paused awkwardly, then added: "Maybe we could even grab a bite to eat while we're out, just you and me?" A peace offering after last night.

I smiled. "That sounds great."

Four days later, on Christmas morning, Jordan and I were sitting on the living room floor, a battlefield strewn with torn and brutalized wrapping paper. She had just opened my present, a necklace with a silver square pendant. Carved into the pendant was an outline of someone swimming through the ocean, head turned in profile, as if taking a breath.

When she flipped it over, her expression turned quizzical. On the back I had engraved a quote. It wasn't an Eleanor line, or even a Dr. Bob original. I'd come across it once while researching fear and had always liked it: *Fear is just excitement without the breath.*

It was an abstract concept for someone her age to grasp—the idea of breath as an antidote to fear. Even I'd been surprised to learn that fear and excitement are biologically nearly identical (think pounding heartbeat, sweating, muscle tension) but that fear can be transformed into excitement by breathing into it fully.

"Holding your breath when you're scared is a way of closing your self off from fear, trying to reject it," Dr. Bob told me once. "But as we know, ignoring fear never works. Instead, inhale and invite fear in eagerly. When you breathe deeply, your anxiety levels lower and feelings of excitement take over."

"I'll explain it later," I said with a wink, and this seemed to satisfy Jordan.

"Thank you!" She ducked her head in that sweetly awkward way characteristic of fourteen-year-old girls.

I went back to plowing through my stocking. My hands touched something flat and crisp, and I pulled out a blank white envelope. Inside was a check made out to me from my dad. I stared at it saucer-eyed for a few moments. My dad was watching from his wingback chair across the room, but when I looked over, he immediately busied himself with his own stocking.

"Your mother and I thought that if you're going to keep on with this . . . *thang* you're doin', you might need some help paying for that mountain of yours," he said, affecting a tone of grudging acceptance. "And we have enough airline miles that you could fly back and forth to Africa for free. We lose the miles if we don't use them so it's only practical . . ."

We all knew full well that there was nothing practical about what I'd been doing for the past half year. And I knew they didn't understand it. And yet, the most practical man I knew was flying me to another continent so I could attempt to climb a mountain, just to see if I could do it. In this instant, I loved my parents so dearly that I was almost in physical pain.

Before I could respond, my mother said, "Can I just say one thing and I swear I won't say another word about it?" She'd been arranging everyone's presents so they'd be opened in an order of escalating delight. But now she came over and put her hand on my arm.

"Just promise me you'll watch out for terrorists," she said. "They'd love to kidnap you and hold you for ransom."

Back in New York, Matt and I decided to spend New Year's with his parents at their beach house. It was there that, at exactly 12:45 A.M., I discovered my bottle of sleeping pills was missing from my tote bag.

"But I *know* I packed them!" I told Matt. "I remember putting them in the bag." I turned the tote upside down in the middle of the carpet. Quarters and lip balm careened across the floor, but otherwise: nothing. Matt paused his flossing to watch me rifle through my suitcase, approaching something close to hysteria.

"Maybe this is a good thing," he offered. "It worries me that you take those pills every night, honey."

I didn't respond. I was too busy strategizing how to get my hands on some sleep meds. Could I make a run to a local pharmacy or drugstore? No, they wouldn't be open on New Year's Eve.

I lay awake for hours, roiling under the covers. Poor Matt suffered in silence next to me although I was sure he wanted to shake me. The plan had been to stay the whole weekend, but I was already plotting which bus I'd take back to the city—and to my prescription sleeping pills—the next afternoon. No way I was enduring another night of this. Then at 4:00 A.M.—as I was contemplating a highly inappropriate raid of his parents' medicine cabinet for a bottle of Nyquil—I sat bolt upright. I knew where they were! I grabbed the keys off the counter, flew out the front door, and had the car trunk flung open in a matter of seconds. Sure enough, the bottle had tumbled out of my purse during the drive. When I made my victorious return to the guest room, Matt had turned on a lamp and was sitting up in bed. I danced the perimeter of the room, shaking the pills around like a maraca. He rubbed his eyes wearily.

"This has become a serious problem, Noelle. You're a drug addict."

I took a sip of water, tipped my head back, and gratefully swallowed a few tablets. "Well, that's a little extreme, don't you think?"

"Extreme? When we were in Aruba, you kept your sleeping pills in the hotel safe with your passport and pearl necklace!"

"You told me to put my valuables in it!"

Matt rolled his eyes and fluffed his pillow a few times before lying down facing away from me. I climbed into bed beside him.

"It's not like I'm getting high," I told his back. No response. "I'm just trying to get to sleep—a basic human function necessary for survival."

"You're taking the easy way out," he said without rolling over. "You need to try harder."

"I need to try harder to lose consciousness?"

"You know what I mean."

"But I *have* to go to sleep. It just takes me longer than everyone else to get there. Imagine if you had a three-hour commute to work, then someone came up with a way to transport you there almost immediately?"

"I'd take the safer option."

"Yeah, that's what everyone says. But I don't see anyone driving a horse and buggy."

"Aren't you worried about what those pills could be doing to your internal organs?" he asked quietly.

In lieu of a response, I turned out the light. The answer was yes, of course, and I'd tried to cut back before. But you'd be amazed and appalled at how easily you'd sell out your liver after a few sleepless nights. I was about to tell him this when I heard his breath deepen. He was fast asleep.

It started at Yale. I learned a lot in college, but the lesson that had stuck with me the longest was how not to be tired. My classmates had graduated from tweedy private schools full of teachers like Eleanor's Madame Souvestre. They arrived at college primed for the rigors of an Ivy League education. They knew how to skim a three-hundred-word book in an hour and retain the information. They cranked out twenty-page research papers in an afternoon and still had time for a game of Ultimate Frisbee before dinner. I, on the other hand, had gone to a high school where only 13 percent of graduates continued their education. The average SAT score had been 876 out of 1600. I'd been as ready for college as Cap'n Crunch was ready to commandeer an actual battleship.

Freshman year, I managed. But the following year my workload increased after I changed my major and had to take extra classes. I was studying until three A.M. just to break even. Over time, I simply trained my body not to recognize tiredness. This was great for studying, but less great when I needed to go to sleep. Sleeping pills were out of the question since I could only afford four or five hours of sleep a night instead of the requisite eight. Besides, the department of undergraduate health wouldn't prescribe sleeping pills to students. I always found this amusing since they handed out free condoms at every turn. If you wanted to sleep *with* someone, they'd supply the provisions, but if you simply wanted to get to sleep, you were on your own.

When my roommate suggested a relaxing glass of red wine before bedtime, I secured a bottle of merlot from the local liquor store that didn't check IDs. Later that night I poured the wine into a plastic tumbler I'd stolen from the dining hall. I took a few sips and made a face. Vile. I hated wine. I walked over to our fireplace mantel where we displayed our liquor, the bottles lined up like trophies. I pulled down a handle of Jack Daniel's and poured myself a shot. Better to get it over with as soon as possible, I reasoned as I tossed it back. Within minutes I felt the booze slip seductively into my veins, my heart rate slowed, and I lapsed into a dreamless sleep. Soon I was having a shot of Jack every night before bed. When one shot stopped getting the job done, I added a second shot. By senior year, my nightcap had progressed to two shots of Everclear, a grain alcohol. At 190 proof, it was more than twice the strength of a shot of whiskey. Yes, it was so strong that it sometimes left me with a sore throat the next day, but having two shots of Everclear before bed made me feel less like a character in a Eugene O'Neill play than drinking four shots of whiskey.

After graduating from college, I moved to New York, where there were plenty of doctors willing to prescribe sleeping pills. I stopped drinking before bed and had many appointments at a center for sleep disorders. They ruled out restless legs and sleep apnea. Physically, there was no reason why I shouldn't be able to sleep. My body eventually built up a tolerance to the pills, just as it had to the nightcaps. Even after taking a sleeping pill, I'd wake up as much as ten times a night. So I added half of a pill. When my body stopped responding to that, I added another half and another . . .

Seven years later, I was up to five pills a night and even Jessica was concerned. "Do all of those pills turn you into a princess or something?" she once asked. "Because otherwise I can't find a logical justification to swallow that much crap. Even a blow job makes more sense."

"It's not like I'm in danger of overdosing," I said defensively. "It takes over forty pills to OD."

When people found out I worked in gossip journalism, they'd sometimes joke, "How do you sleep at night?" "Pills," I'd deadpan. But that was only part of it. Doctors could legally prescribe only thirty sleeping pills a month (because, apparently, they're addictive!), and I'd blow through that prescription in about a week. So I started stepping out on my doctor. At one point I was working two different brands of sleep aids, three different pharmacies, and four different doctors just to get a night's rest. Doctor shopping is illegal, of course, and it was also expensive. It hadn't been as bad when I'd had cushy corporate health insurance benefits, but once I was laid off I'd gotten a cheap plan with a freelancers' union whose coverage policy loosely translated to "Yeah, right!"

The following weekend, Matt spent the night at my apartment. When I came out of the bathroom after washing my face, he was peering at my sleeping pill bottle with suspicion.

"This is a different bottle from last week, isn't it?" he said accusingly.

My guilty expression was all the answer he needed. "I knew it! The pharmacy logo was different. How many of these are you taking a night?"

I couldn't lie to him, so instead I admitted, "Enough that if I keep this up, I won't have enough money to fund the rest of my Year of Fear."

"It's *that* expensive?" he asked.

"You wouldn't believe me if I told you. Also, I can't take sleeping pills on Kilimanjaro."

As soon as we'd finished opening presents on Christmas Day, I'd gone straight to the Internet to read up on the logistics of climbing Kilimanjaro. I'd come across the info about sleeping pills while reading testimonials of hikers who'd conquered the mountain. I'd been heartbroken to learn that taking sleeping pills would be downright dangerous. They suppress your breathing and it's already so suppressed from the lack of oxygen that you could die in your sleep.

It was an impossible situation. I couldn't imagine being that far away from home, in that strenuous an environment, without my pills. What if I didn't get any sleep the entire time? I would never make it to the top, let alone back down again. Yet I couldn't imagine telling my parents I wasn't going. What would I say? That I'd chosen sleeping pills over their generous gift? I couldn't do that, especially not after I'd had that talk with my mom about how she needed to have more faith in me and my sister. If I told her about the pills, she'd have reason to worry about me for the rest of her life.

"You should make facing your addiction part of the project," Matt suggested. "Wean yourself off before you leave for Africa. Is there anything scarier than going off sleeping pills?"

"If there is, I don't want to know about it," I said grimly.

A few days later in Dr. Bob's office, he reminded me, "Research has shown that cognitive therapy is more effective in treating insomnia than sleeping pills."

"I always thought you just said that because you can't prescribe!" Dr. Bob was a PhD, not an MD. "Like the guy with the tiny penis who says, 'Size doesn't matter!' "

He gave me a warning look.

"Sorry."

"Most insomnia is due to excessive mental activity—namely, worrying," he said. "Tell me, what do you think about now when you're trying to fall asleep?"

"Lots of stuff. What if I can't make rent? What am I going to do for a living when this year is over? Is Matt 'The One'? Is that Ben Affleck's real hair?" I paused, realizing Dr. Bob might not know who Ben Affleck was, but he was already jumping in.

"First understand that we're hardwired to toss and turn all night," he explained. "Anxiety helped our ancestors survive in a primitive environment. In circumstances where animal attacks could happen at any moment, where strangers could kill you, where your survival could depend on whether your tribe liked having you around, those who weren't anxious enough didn't survive."

"But why am I worrying now? We don't live in that kind of world anymore."

"Precisely. Modern civilization is eliminating most of these threats far too rapidly for our evolutionary biology to catch up," he said. "In fact, rates of anxiety have increased dramatically during the last fifty years. The average *child* today exhibits the same level of anxiety as the average psychiatric patient in the 1950s."

"And these are the people who are going to be running the country in forty years?" I said. "That's reassuring."

But he had a faraway look in his eye. "Now most of our worry is unproductive worry. We worry about past mistakes, obsess about what other people think of us, create terrifying future scenarios out of nothing. Our mind chatters away even when we wish to sleep or relax or simply do nothing."

I was reminded of an Eleanor quote I'd come across a few weeks before but hadn't understood: "Most of us, I suppose, are ridden by at least some imaginary fears. But I think it is as important to deal with these as it is with the fears based on a reasonable foundation. They often do us more harm." Now it made sense to me.

Dr. Bob snapped out of his reverie. "The good news is, by working to overcome your anxiety, you've taken the first step toward overcoming insomnia," he said. "Now it's time to go all in."

"All in?" I asked nervously.

"You have to make a choice," he said. "Are you going to continue on this path or change direction? What happens when you start taking three pills a night? Four?"

It seemed unwise to tell him I was already taking five.

He continued: "Sleeping pills artificially alter your circadian rhythms. In order to beat your insomnia, you're going to have to get off the pills."

A surge of panic coursed through me. Insomnia made me feel like a prisoner of my mind. Lying in the dark, I had nothing to do for hours but think. I was trapped with my worries. Pills were my only escape from that prison, a way to escape myself. Wasn't I already facing enough fears this year without taking on this as well? Now I had to face fears at night as well as during the day? There was no way I could do it. Not now. Maybe next year, when my project was over and I'd found another job and I settled back into some semblance of a normal life, then I could focus on tackling this issue. But now? Was he serious?

Chapter Nine

⁓

Happiness is not a goal, it is a by-product . . . For what keeps our interest in life and makes us look forward to tomorrow is giving pleasure to other people.

—ELEANOR ROOSEVELT

And then the hospital called. It had been more than a month since I'd sent in my application. I'd assumed they'd done a background check and discovered I'd been arrested in college for talking back to a cop who raided our party and decided I was a security risk. Instead they told me they loved my essay and had an opening in their volunteer program.

After an interview, an orientation, and extensive medical and drug tests, I was placed as a volunteer with the Milkshake Program, making milkshakes for the oncology ward of the hospital. My co-volunteer was an adorable twenty-three-year-old former gymnast/cheerleader named Becca, who had freckles sprinkled across her nose and was terrifyingly cheerful. Despite all this, I liked her immediately. Our assignment was simple: go from room to room asking if patients wanted chocolate or vanilla, mix the milkshakes, deliver them. The oncology floor was set up like a dorm. Rooms lined a long hallway, each room with its own front door and private bathroom. Many patients had a roommate, the two beds divided by a curtain.

On my first day, I hung back and watched Becca take a few orders. She was applying to medical school and volunteered in various wards throughout the hospital, so she was accustomed to dealing with patients. At first my voice sounded high, practically falsetto. I spoke in that exaggeratedly nice way that women speak to other people's children. But after the first few patients, I relaxed and found my groove.

Most of our time was spent preventing ourselves from infecting patients. During orientation, I learned that one hundred thousand people die every year from infections they contract at a hospital, so before someone could enter a patient's room, precautions had to be taken. First, patients were assigned to one of three categories to indicate the fragility of their immune system—Hand Hygiene, Contact Isolation, and Droplet Isolation—which was posted outside their door to signal what needed to be done before entering. Hand Hygiene meant coating your hands in sanitizer. Contact Isolation meant coating your hands in sanitizer, putting on rubber gloves, and donning a long-sleeved gown over your clothes. Droplet Isolation meant sanitizer, gloves, gown, and a dense surgical mask that made your breath come out in a dull roar and had a clear plastic window that stretched up over your eyes like a personal windshield. The whole effect was slightly Darth Vader. These rooms also had a negative pressure "anteroom" between the hospital hallway and the patient's quarters with a special ventilation system designed to prevent airborne infections from traveling in or out. You had to be sure to close one door completely before you opened the other. After leaving each patient's room, you did The Dump, where you undid everything you'd done a few minutes ago: throw the gloves and/ or mask in the garbage, toss your gown in a hamper, and apply hand sanitizer again. Then you moved to the next room and started over. There were sixty rooms in all.

Back in the milkshake-making room, I shook my head in wonder. "It just seems crazy to risk someone's immune system over a milkshake."

Becca snapped on a pair of gloves. "The benefits are twofold for the patients. They need the calories, and it's a morale booster. Don't underestimate the power of a milkshake." She handed me the ice cream scooper. "So what did you think?"

I dragged the scoop over the ice cream so that it rolled into a vanilla wave. "It's not how I thought it would be." What *had* I expected? Movie cancer? For every patient to be pale and bald, surrounded by family members who'd shaved their heads in solidarity? "A lot of them look so . . . *normal*. I'd never know some of these people are sick."

"It's the treatment that causes the baldness, not the disease," Becca shouted over the whir of the blender. "Chemotherapy kills all cells—healthy cells just repair themselves faster than cancer cells. So it kills the cancer slightly faster than it kills you."

Later that night, I popped my usual five sleeping pills and went to bed. I'd run out of the brand of pills I preferred, which knocked me out immediately. The backups took longer to kick in and were known to cause hallucinations and short-term amnesia. Sometimes I'd take the pills and a few minutes later I'd forget that I'd taken them. Then I'd have to dump out the entire bottle and count the pills backward from the date the prescription had been filled.

Tonight it was just hallucinations. It was my stuffed animals coming to life that clued me in. I lay in bed watching them move next to me, their arms and legs shifting as if they were trying to find a better sleeping position. The first time it happened, I was in the bathroom doing my final pee of the evening when I looked down at the *Us Weekly* on the floor. The stars in the photos were waving at me. Convinced I was imagining things, I returned to my bed, but I stopped short of getting in. It looked like someone was hiding under my comforter breathing. I could actually see the blanket rising up and down. With a tentative hand, I jerked it back, but nothing was there.

By now I was used to this. I simply rolled over away from the stuffed animals. I put on a sleep mask so I wouldn't see the shadows in the room moving on their own, Peter Pan–style. By far the freakiest side effect of the drug was the feeling that there were other people in the room. One time during a conversation with a friend who took the same pills, I mentioned the moment "when the people come," and his face lit up with recognition. He knew exactly what I was talking about. Sometimes they talked to you and seemed so real that you actually talked back. The weird part was that it didn't seem weird at the time. My whole life I'd worried about people breaking into my room at night. Now for all I knew, there were people in there, but I wasn't afraid. And *that* worried me.

Even more than milkshakes, the patients seemed to crave normalcy. Most of their human interaction was about their illness. So I acted as though we weren't in a hospital but in a restaurant somewhere and I was taking their dessert order. If the patient seemed up to it, I played the role of the sassy waitress. I'd wait while they debated flavor options. This was the tastiest thing they'd eat all week, so they took the decision seriously.

"What do you recommend—vanilla or chocolate? Or should I go half and half? I don't know!"

I'd lean in conspiratorially. "I mean, we all know chocolate is the superior flavor, am I right?" Or I'd say, "Let's go crazy. Half and half." If they asked for strawberry, I pretended to balk. "*Strawberry?* We don't have strawberry. What are you—some kind of health food nut? If it's fruit you're after, maybe the nurse can find you a fruit cup. Now are you ready to get down to business?" They loved it.

I was humbled thinking of the lengths Eleanor went to in order to visit wounded soldiers during World War II. Namely, cramming her fifty-eight-year-old body into a small bomber plane notorious for its tendency to catch fire, then traveling twenty-three thousand miles to Australia, New Zealand, and seventeen South Pacific islands. The five-week trip was brutal. Along the way she lost thirty pounds. Opponents dismissed her goodwill tour as a publicity stunt and criticized her for flying around the world on the government's dime. Nevertheless, she worked from sunrise to sunset, driving hundreds of miles between hospitals and camps, meeting more than four hundred thousand troops. She toured countless hospitals, stopping at every bed in every ward, chatting with wounded servicemen at length. Sometimes their injuries were so severe Eleanor had to keep herself from flinching in front of the patients.

I was learning to steel myself, too, before walking into each room. Some patients were painfully thin. Sometimes they were missing body parts. There was a man who'd had all the toes on one foot amputated. He was asleep when I delivered his milkshake, his foot propped up via a pillow at the end of the bed. It looked so anonymous, like a head without a face. All of the character of the foot is in the toes, I realized.

One week I walked into the room of one of my regulars. "Hey, Doris! What are you feeling today—chocolate or vanilla?" I asked but only heard wheezing in response. I looked up and saw she had a hole in her throat with a tube sticking out of it. She'd had a tracheotomy and couldn't speak. I thought for a minute. I couldn't give her my pen to write with because it was covered in germs.

Taking care to keep my voice warm and level, I said, "Okay, Doris, how about we play the thumbs-up, thumbs-down game? Thumbs-up if the answer is yes, down if the answer is no. Do you want a milkshake?"

Thumbs-up.

"Do you want chocolate?"

Sideways thumb. What did that mean? She mouthed something that I couldn't quite make out.

"Er, so you want vanilla then?"

Another sideways thumb. Understanding dawned. "Oh! Do you want a mixture of the two?"

She nodded happily. Double thumbs-up.

Sometimes there was singing. One week it came from the room of an emaciated woman in her forties, wearing a brightly colored scarf over her head. When I'd stopped by her room earlier, she'd been unconscious, breathing lightly, unaware of the loved ones taking turns holding her hand. Now someone had brought out a guitar and strains of "Que Sera, Sera" spilled into the hallway. I leaned against the wall outside the door for a few minutes, listening. I felt privileged to witness such poignant scenes, but I also felt guilty trespassing on someone's private moment. Then Becca passed by and, in a sad voice, whispered exactly what I was thinking: "That sounds like a swan song."

Down the hall was Mr. Orth, who had become one of my favorite patients. He was a sixtysomething, balding (not cancer balding, but balding balding) former tax attorney and practicing shameless flirt. When I opened his door to deliver his milkshake, he was chatting on his cell phone but waved me in.

"It's just my brother," he said, palm covering the mouthpiece. Then he switched the cell to speakerphone and set it on the table so he could unwrap his straw with both hands. "Hang on," he said loudly, "there's a candy striper here."

"Is she a hottie?" the voice on the phone asked, clearly not realizing he was on speakerphone.

"That depends," I called out. "How hairy do you like your women? If you're into the Chewbacca thing, I'm your girl."

Mr. Orth hooted and asked, "Say, Candy, are you married?"

Out of instinct, I almost said: *Are you crazy? I'm not old enough to be married!* I was, of course. Yet the question seemed as inconceivable to me now as when I was a child and the school photographer would ask that to make me laugh and show my teeth.

"Why?" I asked. "You looking for a tax break?"

"Do you have a white dress?" Mr. Orth pressed.

I looked down at my slacks and volunteer smock. "Not on me."

"How about you find one, I put on a suit, and we run off and get married?" he said with a wink. Honestly, had he not already been married, it would've been the sweetest proposal a girl could hope for.

I put my hands on my hips and nodded at his hospital gown. "Well, you're already wearing a dress. Why can't I wear the suit?"

There was silence for a few beats and I worried I'd gone too far. Then Mr. Orth burst out laughing, and so did his brother on speakerphone.

When I stopped by to see Mr. Weiderstein, a man in his eighties, he was surrounded by all six of his boisterous children.

"Come on in! The water's fine!" crowed his son, a bearded heavyset man in his fifties. You'd think they were crowded around a table at a dinner party the way they were carrying on, though their father was lying there bedridden, and every time he tried to speak, he could only gasp. Mr. W. was scheduled for brain surgery the next day but he could have solid food until midnight.

"So bring him a large vanilla," his son winked. "Go big or go home—that's our motto."

"A wild man! I like it!" I nodded at Mr. Weiderstein with approval and scribbled the order on my pad. The group waved cheerfully as I left the room.

A few minutes later I was still standing in the hallway at an isolation cart struggling to pull on a pair of gloves when the son walked slowly past me.

"Oh, I meant to ask you," I said, "who is your father's nurse?"

He looked at me with irritation and said curtly, "I don't know. Whoever it says on the bulletin board." I turned away, a little stung. A minute or so later I heard great gulping sobs. When I turned back, he was leaning against the wall, hunched over with tears pouring down his face.

"Oh God," he whimpered. "Oh God."

Oh God. Should I say something? There was no official hospital policy on this. It was a judgment call. When he'd spoken to me a second ago, he seemed like he wanted to be left alone. So I quietly closed the drawer to the isolation cart and padded away. Before I turned the corner I glanced back and saw his sister standing next to him, holding him in sort of a sideways hug. I kept going, his moans following me down the hallway.

During her fourteen years as First Lady, Eleanor received an average of 175,000 letters a year. One such letter was from a destitute young woman named Bertha Brodsky, who apologized for her handwriting, explaining that she was bent sideways due to a crooked back. Eleanor set Bertha up with a specialist, arranged her surgery, visited her in the hospital, and sent gifts on holidays. After Bertha recovered, Eleanor helped her find a job, attended her wedding, and became godmother to her child. The woman was nothing if not thorough. I had hoped for my own stories like this. But since all I could do for the patients was bring dessert, it was hard to feel like I was really making a difference. Sometimes I felt I was actually making things worse. Those are the moments I remember the most.

One day I was taking a patient's milkshake order and her roommate, who was a diabetic, overheard.

"Can I have one too?" a voice called plaintively.

I walked over to the other side of the curtain separating the beds. An Asian woman in her forties looked at me with pleading eyes. "I'm sorry, but your blood sugar is too high this week," I said gently. "I'm not allowed to give you one."

"Please? I won't tell anyone."

I tried to fathom how sick you'd have to be to be willing to make yourself sicker just for a moment of relief. "I'm sorry, but I can't."

She burst into tears. "You don't understand!" she said, burying her face in her hands. "This has just been the most awful day."

I'm a monster, I thought.

Becca and I always took the subway home together. When we got to her station, she hopped off and called out, "See you next week!" I reached into my backpack and pulled out the *New York Times*. At the next stop, a group of kids clattered onto the train, talking so loudly that I looked up from my newspaper. They all wore some kind of school uniform, but two of the boys also had baseball caps. The girls had rolled the waistbands of their plaid skirts to raise the hems to horrifying heights. Another boy was with them, and though he wore the same clothes and haircut, he didn't quite fit. There was an air of desperation around him, an overeager quality that teens, especially, seem to find unappealing. When he tried to sit down with them, one of the baseball cap boys stretched his leg out across the remaining available seats.

"Uh, did we say you could sit with us, loser?" he sneered.

The kid, visibly deflated, took a seat a few sections away, directly across the aisle from me. He stared very hard at the subway ad over my shoulder, on the precipice of tears.

There's an unspoken rule among New Yorkers that you don't speak to people you don't know on the subway. But something about this boy pulled at my insides. I wanted to build a time machine just for him so I could show him that in a few years, he wouldn't care what these idiots thought. Also, I was still questioning whether I'd made the right call that day with Mr. Weiderstein's son, and I didn't want to regret not saying anything to the boy. I leaned forward in my seat, resting my elbows on my knees. I wanted to be close enough that the other kids wouldn't hear.

"It won't always be this way, I promise," I told the boy.

He looked at me, startled, clearly wondering if I was a subway weirdo or someone who wanted to tell him about Jesus. The people near us looked at me oddly too. I could feel my cheeks tingle a little but didn't care.

Holding his gaze, I said: "People are meaner when they're younger. They just are. But it won't always be this way. Once you're out of high school, you'll be surrounded by all new people. And almost everyone you meet will be nice to you. This world that you're in right now, it's not the real world. Just remember that. One day there will be lots of people who'll want you to sit with them."

I didn't want to make him uncomfortable by waiting for a response, so I went back to reading. A few minutes later, I sneaked a glance over the top of the paper. He was staring off into space again, but with a small smile on his face.

Mrs. Andrews was in her late fifties. Over the last two months I'd watched her beautiful red hair dwindle to a few select strands. When I'd asked for her order last week, I could tell it had taken all her effort just to listen to what I was saying.

This week I entered Mrs. Andrews's room cautiously, but to my relief, she was still there. Barely. Even in her sleep, she was breathing laboriously. A man I assumed was her husband was folded into a chair in the corner. He was rumpled, like he'd been sitting there for days.

Rubbing his hands over his haggard face, he said, "Her condition just keeps getting worse, Doctor. Isn't there anything you can do?"

I looked over my shoulder, but no one was behind me. He thought I was a doctor. In the Contact Isolation rooms, all hospital staff wore the same long-sleeved yellow gowns, so it was difficult to tell if we were wearing a volunteer apron or a doctor's coat underneath. His eyes were desperate, looking for hope. I couldn't give him hope. I couldn't even give him strawberry.

"I'm sorry, I'm just a volunteer. I, uh—I make milkshakes." I said, feeling ridiculous. "I'm here to take your wife's dessert order. Do you happen to know what flavor she prefers?"

He blinked at me for a moment, confused. "What?"

"I'm taking milkshake orders. We have vanilla or chocolate . . ." I trailed off.

His eyes went down to his lap. He didn't respond.

Finally I asked, "Can I help you with anything else?" though it was clear I couldn't.

Visitors rarely said the right thing when leaving a cancer patient's room. One day I delivered a milkshake to a patient just as a group of his work colleagues was leaving.

"Get well soon!" a few of them said.

One of them patted the man's spindly shoulder. "You'll be back at the office in no time!"

Their remarks were patently absurd. A few months ago I'd have said the same thing. Now I wanted to shake them. The man smiled indulgently at his friends but he obviously knew—as we all knew—that he'd never set foot in an office again. I'd recently overheard one doctor telling another that he was DNR—Do Not Resuscitate.

"Usually it's because they're terminal," Becca explained later. "The patient is basically saying, 'Let me die.'"

How isolating it must have been to have your loved ones faking it all the time. To be *handled*. To be spoken to like a child whose parents insisted on pretending there was a Santa Claus long after you'd stopped believing.

I wanted to reach out in moments like that, but it was inappropriate, given my position. There were counselors here who came around to talk to patients about such matters. And Becca and I only got a few minutes of interaction with each person, hardly long enough to become a confidante. But there was something else I could do for them, I realized. These people's organs were being ravaged by disease and I was willingly damaging my own every night. Out of respect for them, I knew I had to give up the sleeping pills.

Chapter Ten

Yesterday is history. Tomorrow is a mystery. Today
is a gift. That's why it is called the present.

—ELEANOR ROOSEVELT

This was easier said than done and, frankly, not that easy to say. In the two weeks that followed, I told no one about my decision because I was afraid I would fail and then I'd have to face their disappointment. I didn't even tell Dr. Bob. During our next session when he asked what I wanted to talk about, I brought up something else that happened that week. I'd been researching a freelance article about Internet dating when I came across the personal ad of a woman named Jasmine. In response to the question *If you could be anywhere in the world at this moment, where would you be?* Jasmine had written, "In the moment." Most of her competition had gone with Cancun, so her answer stood out. *How often am I really in the moment?* I'd wondered.

"Well, how often *are* you in the moment?" Dr. Bob prompted.

"I'm not even sure I know what that means. That's why I asked you," I said with more edge to my voice than I intended. Two weeks of sleep deprivation was making me grouchy.

He didn't acknowledge the change in my tone. Instead he took a deep breath, the kind people take when they're about to explain something that's fairly complicated. "What we're really talking about here is mindfulness," he said. "Mindfulness is a technique where you concentrate on the immediate present experience, without judging or trying to control what is going on. To be fully aware. This practice is at the heart of many forms of Eastern meditation, especially Buddhist forms. But you don't have to be Buddhist to practice mindfulness."

"So you're suggesting I take up meditation?" I was already spending hours a night lying still with my eyes closed, alone with my thoughts. The idea of doing that on purpose was unbearable. "That doesn't fit into my project. There's nothing scary about sitting cross-legged and chanting for a while," I snapped.

His face remained patient, but he clearly knew something was up. So he waited me out. Finally I blurted out, "I'm sorry. I'm just so tired. I'm trying to get off the pills." I was reducing my intake gradually, because going off five pills at once was dangerous. Still, going down to four pills a night kept me tossing and turning for hours.

"This is great news," he said warmly. "Not that you're tired, of course. But it actually speaks to what we're talking about. We know that your insomnia is linked to anxiety. Practicing mindfulness will help you address the root causes of your worry."

Dr. Bob propped an elbow casually on the arm of his chair. "Not only that," he went on, "mindfulness will help you stay in the present, where fear does not exist. Fear exists in the past, like worrying about the dumb thing you said to your boss yesterday, or in the future, as in fretting over whether your plane will crash."

I tried to imagine living in the present. As a blogger, I'd been trained to live in the future, always looking for the next story. The day Tom Cruise and Katie Holmes's baby was born, my editor had stopped by my desk and asked, "What are you working on?"

"Well, Suri Cruise was born this afternoon—" I started.

"That story is so one-hour ago. Get Katie's trainer on the phone and write a story on how she plans to lose the baby weight. Then round up some photos of every celebrity baby born in the last three months and let's do a poll asking, 'Who do you think Suri Cruise should go on her first playdate with?'"

"The kid hasn't had her first bowel movement, and we're booking her social calendar?"

"We have to keep advancing the story. Always be asking yourself, 'What's next? What's next?'" the editor said, snapping her fingers together in rapid succession. That was how I'd lived most of my life, actually. In high school, I'd been focused on getting into college. At Yale, everyone was obsessed with landing the best summer internships. After graduation, I turned my attention to getting a job, then getting another job, then a promotion, and so on.

"But how does that help me manage my fears?"

"We tend to treat our thoughts as though they are reality. If we think something, then it *is* so. We tell ourselves *I'm a failure* or *My life is a mess* and accept it as truth, and our emotions get worked up about it." Dr. Bob leaned forward. His grayish brown curls did a little dance and settled into their new position. "Mindfulness teaches us to view our thoughts as *just thoughts,* not facts. We don't have to be afraid just because we think fearful thoughts."

"But I've tried meditating in yoga class, and my mind always ends up getting distracted."

He gave me a reassuring nod. "You might start by paring down the distractions in your life so you can focus more on the moment. Television, for example, is used to escape real life. Cell phones and the Internet claim to make us more connected but actually pull us away from real relationships. And they make it difficult to connect with ourselves."

"Giving up all my technology, now that would be scary," I said, half in jest. Then my mind flashed back to August 2003. There had been a massive blackout in the Northeast, and New York lost power for two days. People were trapped underground on subways and in elevators that had stopped without warning. All stoplights went dark. Tourists flooded the streets, unable to open their hotel room doors with electronic keys. People had no money but what was in their wallets because ATMs didn't work. The governor declared a state of emergency. Yet the hardest part of the experience for me had not been climbing thirty flights of stairs to my apartment or not having air-conditioning in stifling heat. It was the restlessness and loneliness that came from not having cell-phone reception, Internet, e-mail, or TV.

"First time on Cape Cod?" my cabdriver asked as he pulled away from the Hyannis bus station.

"It is! I mean, I've been to Nantucket before but—"

"Not the same," he said crisply. "Here on business or pleasure?" His Kennedy-esque accent was fantastic.

"Both, I guess. I'm here on a five-day silent retreat."

"You're paying someone else so that you don't gotta talk? Hey, whatever works for ya."

"It's more than just not talking. I'm cutting off the Internet, e-mail, texting, the works."

"Don't own a computer, don't wanna!" he said proudly. Based on the third of his face visible in the rearview mirror, I placed this guy somewhere in his eighties.

He gestured toward the distance. "Down thaddaway is Kalmus Beach. Named after the fella who invented Technicolor," he said, pronouncing it *Tek-nee-kuh-luh*.

It was drizzling heavily, and I saw the world as if through a screen door. The houses were, unsurprisingly, Cape Cod style with steep roofs designed to easily shrug off the snow, shingles weathered to a distinguished gray that matched the color of the sky.

"Say, is it usually in the forties in April?" When I'd booked this trip, I'd pictured myself taking daily walks on the beach, feeling the warmth of spring on my back.

"Nippier than usual out here this week," he said. "The cold off Nantucket Sound just seeps into ya bones.

"See that marsh out there to our left?" he asked, nodding his head at the radiant crimson swamp that could've been designed by Mr. Technicolor himself. "That's a cranberry bog."

"No kidding!" I craned my neck. It was like being in the presence of celebrity after all the vodka cranberries I'd drunk in my life. I continued gawking until it dropped out of sight when we turned off into a neighborhood.

Most silent retreat centers are religion-based, so since I'm a practicing Catholic, I'd gone with the closest Christian one I could find. I attended church every Sunday, which used to instill a sense of peace that lasted for the rest of the week. But lately, sitting in the same place for an hour felt torturous. I was on my knees like everyone else, but instead of praying I was checking my watch, thinking, "Can we move this along? I was hoping to get out of here before the Second Coming of Christ."

"Ah, here we are," said the cabbie. He stopped in front of a house with a sign in the front yard announcing the name of the retreat. There were about ten small cabins flanking the driveway. As I scooted out of the car he handed me his business card. "In case you get tired of the silence," he said with a wink.

"What is it that you're looking to release here and what are you trying to receive?" asked Alice, the fortysomething woman who ran the retreat center. We were sitting in the country-style dining room to fill out some last-minute paperwork. She was one of those incredibly serene people who've either answered a religious calling or stashed bodies in their basement. She had a shoulder-length mom haircut she was allowing to gracefully turn gray. Behind her rimless glasses, her face was completely unlined.

"I'm afraid I'm turning into one of those people who can't live without their cell phone, TV, or the Internet," I admitted. "So I guess I want to release all those distractions and relearn how to just . . . *be*."

"Silence is about quieting your mind as well as your voice," she said, pushing a lock of her sensible hair over one ear. "Most people find it difficult to be truly be alone with their thoughts. Would you like me to teach you a technique you can do on your own to calm your mind?"

"Sure."

Positioning her hands over her heart, she closed her eyes and said, "I am here." She waited a few beats and opened her eyes.

I cocked my head. "That's it?"

"Simple as that. You're telling your heart that you're present and ready to listen to what it has to tell you."

She handed me a pen and a form to fill out my address and credit card information. "You're free to do as you please here. The only thing we ask of our guests is that they respect one another's silence by not speaking to them—though you are allowed to sing during services. From now on I'll speak to you only when necessary."

I nodded, unsure whether I was supposed to be silent already. Alice pointed me toward one of the small one-room cabins flanking the driveway. When I walked in, I peeled off my coat but quickly pulled it back on with a shudder. The thermostat read sixty-seven. I cranked it up to eighty.

You think this *is cold?* I thought to myself. Exactly three months from today I'd be climbing Mount Kilimanjaro, where temperatures could reach thirty below zero at the peak.

I knelt over the roaring fireplace in the corner. Feeling no warmth on my hands, I peered inside. The flames were fake. So were the logs, which were also—gloriously—coated in glitter.

Smiling, I stood up and noticed the wicker bureau for the first time. Sitting on top was a new flat-screen TV. Next to it, a cheerful sign announced that complimentary wireless Internet was also available. I'd forgotten that the retreat operated as a regular hotel during tourist season. I pulled out my cell phone and made one final call before going silent.

"So I think my fireplace is gay," I announced when Matt answered.

"What?" The Albany newsroom was abuzz in the background.

"Never mind." I flopped down on the faded floral comforter. "Just wanted to let you know I got here okay."

"How is it?"

"It's a little unsettling, to be honest," I said, looking around my cabin. One of the walls, I noted, was a sliding glass door with a flimsy lock. "It feels like one of those horror movies where they lure you here to experiment on you, knowing no one expects to hear from you for five days. And by the time people catch on, my captors will have already fled with my internal organs."

"I can't believe I don't get to hear your voice for five days," he said. Matt and I had never been one of those couples who fight and then take a few days to cool off. We basically never fought at all. Our longest argument took place on the F train and we didn't speak for three subway stops.

"Me too," I said, swallowing back the lump rising in my throat. It was hard enough not seeing his face five days out of the week.

After we hung up, I reluctantly shut down my phone. I set the BlackBerry on the bureau and stared at the forbidden electronics. Not exactly the three temptations of Christ, but having them in the same room for five days would make them harder to resist.

There was a knock at the sliding glass door.

Alice poked her head in. "We're having a community prayer service in a few minutes, which you're welcome to attend."

I nodded, eager for a change of venue. I put on slacks and followed the brick sidewalk to St. Mary of Magdalena Chapel, a handsome cottage located behind the main house. As I slipped inside, Alice materialized and handed me a prayer book.

"We're so pleased you could join us. Right this way." There were roughly thirty chairs with hymnals placed on the seats. In the middle of the front row sat a heavyset woman in her midfifties, wearing spectacles and a baggy sweater.

"This is Margaret. She arrived yesterday," Alice said. Margaret and I smiled in silent greeting. "Shall we get started?"

My smile froze. *Wait, what? I thought this was a community prayer service. Three people isn't "a community." It's not even doubles tennis.*

The way Alice explained it, this was a call-and-response with Margaret and me singing alternate verses. It had been only a few months ago that I'd first mustered the courage to sing in front of other people. That had been a rap song with a booming sound track, and this was two people going a cappella. Margaret looked over expectantly as she opened her songbook. I was about to politely decline, but then I remembered I was not allowed to converse. And so commenced an incredibly awkward twenty minutes of my struggling to read music. Trying to hit the high notes felt like trying to knock a cereal box off the highest shelf in a grocery store. It was a relief when Alice moved on to the prayer readings.

Shortly after the service, it was time for dinner, which was served at 6:00 P.M. every day in the kitchen of the main house. Margaret greeted me with a head nod when I walked in. It turned out that the reason why the prayer service had been so sparsely attended was because she was the only other guest this week. A moment later Alice pushed through the swinging kitchen door, back first, holding a tray. On it was a dish of red beans from a can, populated with pieces of hot dog. Dessert was a loaf of instant cornbread pockmarked with blueberries. She set down a bottle of ketchup and said, "Stack the dishes on the counter when you're done. Peace be with you." Then she took her leave.

Margaret bowed her head silently. I bowed my head, too, though I didn't normally say grace before meals. Instead, I silently told the beans, *I am going to regret you later.* Sure enough, by the end of dinner, Margaret and I sat in mutual embarrassment as our stomachs called out like gastrointestinal whale songs. Hers emitted high squeals and mine answered with low rumbles. The eating itself had been awkward.

To keep from staring at each other, we looked down at our plates or around the room, pretending to be interested in the kitchen appliances. It was like being on a bad date. As always happens in inappropriate situations, a voice in my head urged me to do the "Ooga Booga Pee Paw!" thing. Or what if I leaned over and squeezed her boobs while shouting, "Honk! Honk!" Would she break her vow of silence and react? Keep attending to her beans? The mind reeled. When it was over, we hastily stacked our dirty dishes on the tray and repaired to our cabins. The clock told me only fifteen minutes had gone by since I'd left. It was amazing how quickly dinner passed without conversation or a sitcom to linger over.

My room was still freezing when I got back so I drew a bath. I wouldn't have figured Alice to be the type to install a Jacuzzi, but I liked where her head was at. I cranked the water as hot as it would go and stepped delicately into the tub. My scalded skin bloomed into a happy pink. Leaning my head against the wall behind me, I said a quick prayer of thanks, wondering if I was the only one whose prayers sounded like an Academy Award speech. "I'd like to thank God for my Jacuzzi. Really, this was just so unexpected. I don't know what else to say. Um . . . free Tibet?"

I closed my eyes and wondered what I should do tomorrow. Maybe walk the labyrinth behind the main house? When I'd first read about it on the retreat center's website, I'd pictured a maze with towering walls or people-eating bushes that magically rearranged themselves like in the movies. But as I'd passed it on the way to the chapel earlier, I saw it was just a brick footpath on the ground. Still, in expectation of walking this path, I'd looked up the history of labyrinths.

During the early Middle Ages, there was a cathedral boom across Europe. Twenty-two cathedrals were built with labyrinths placed in their floors. Up until then, Christian pilgrimage to Jerusalem was regarded as a sacred obligation of faith. But when the Crusades swept across Europe, the journey became dangerous and often deadly. The Christian leaders came to the same conclusion that all smart kids reach when they dare to venture into a bad neighborhood at night: "Uh, let's not and say we did." So it was decided that walking a labyrinth could serve as a substitute pilgrimage for the faithful unable to travel to the Holy Land.

Over time, walking the labyrinth became a metaphor: the path to the center symbolizes the journey inward to our own center. One loses track of direction in its twists and turns, and theoretically, the distractions and anxieties of the outside world are left behind. This quieting of the mind allows you to open yourself to the journey. The center of the labyrinth represents the ultimate surrender of self so that one can receive peace, clarity, illumination, or God. When I'd read that, I was skeptical but intrigued.

New Age mystic Jean Houston was considered the grandmother of the modern labyrinth revival, even adopting a labyrinth as her logo. In the 1960s, she began advocating walking their coiled paths as a means to spiritual enlightenment. She acted as something of a guru to Hillary Clinton when she was First Lady, leading her into guided meditation sessions to contact—wait for it—Eleanor Roosevelt. She had Clinton carry on imaginary dialogues with Eleanor, with Clinton supplying both sides of the conversation.

My thoughts were brought back to the present by the water, which had turned lukewarm. Had my body already gotten used to it? I'd only been in here a few minutes. I plunged my right foot under the still-running faucet. Ice cold. My good mood turned instantly. I splashed my two inches of tepid water around in frustration.

"Oh, come on!" I called out. "Seriously?"

The combination of franks, beans, and tapering off my sleeping pills kept me up half the night, so I let myself sleep through the continental breakfast. By lunchtime I was ravenous and arrived to find the dining room empty. There was a note on the counter from Alice:

Margaret + Noelle,
   I left butternut squash soup and a chicken salad sandwich in the
refrigerator.

Enjoy!
—Alice

I peered inside the fridge. There were two Tupperware cups full of soup and half of a small chicken salad sandwich on a plate. Margaret must have come earlier, eaten her half of the sandwich and passed on the butternut squash. I popped the container into the microwave for a few minutes and settled down at the table with my portion. As I was sampling the soup, trying to determine if it was actually curry that had been repurposed, Margaret walked in. She opened the fridge and cocked her head. She read the note and looked over at me sitting at the table with my food. She opened the fridge again. Margaret hadn't been here already, I realized. Confronted with the options of entering into a lively game of charades or breaking our vow of silence, I went with the latter.

Hoping she wouldn't be too offended by the intrusion, I said, "This is all there was." My voice was a little too loud and reverberated across the linoleum.

She looked at me gratefully. "Oh, thank you, I was so confused!"

"We can split this," I offered, pushing the plate with the sandwich into the middle of the table.

"I appreciate it!" After heating up her soup, she sat down across from me, picked up the plate with the sandwich, and set it directly in front of her.

Oh, this is uncomfortable. Maybe I should just let her have it and not say anything? We're having such a nice moment. Why ruin it? But Dr. Bob would call this "avoidance."

I cleared my throat. "Um, actually, when I said that this is all there was, I meant that this half sandwich is all there was. I only just got here myself."

"Oh!" She flushed red. "I do apologize! I thought you'd already eaten one half."

"You can have it if you want," I said, suddenly feeling generous in spite of my hunger.

"Nonsense! We'll divide it." She reached for a butter knife and began the dissection. "I saw a teenager working in the yard earlier. Perhaps Alice offered him a sandwich and he took more than his share? Boys eat a lot, you know."

"You have to give him credit for a brilliant strategy: steal food from the silent retreat guests. Who are they going to tell?"

She returned my grin, and we munched our quarter sandwiches and sipped our curry/soup with a new air of comradeship. Miraculously, when we finished I was completely full. It was exactly enough.

Margaret's maternalness made me miss my mom. I wished I hadn't been so hard on her at Christmas. The more I'd learned about worry in the last few months, the better I understood how easily worrying became an addiction. People worry for many reasons, according to Dr. Bob. We believe that if we chew over a problem long enough, eventually we'll figure out a solution. Worry gives us the illusion of control over the future. We dream up worst-case scenarios, thinking we can prevent bad things from happening. We think worrying will motivate us to get things done. We worry about exams, thinking it will get us to study. We worry about our appearance, hoping it will encourage us to work out or stick to our diet. Also—and this was a hard one for me to wrap my head around—we worry because it makes us less afraid.

"Worry is your body's way of trying to suppress fear," Dr. Bob had explained in our last session.

"Aren't fear and worry basically the same thing?"

He'd shook his head. "Fear is an emotional response. It manifests physically. Think tension, muscle aches, rapid heartbeat, sweating. Worry suppresses that arousal."

"So it's more like a defense mechanism."

"It temporarily makes us feel better, so we keep doing it."

ے

It continued to rain through the next day. Even an umbrella was use-
less because the cold wind whipped the drops around so they came at
you from below, delivering uppercuts of sogginess. Thankful that I
had brought a few Eleanor books along, I spent the day reading her
thoughts on religion. All I knew was that she was a lifelong Episcopa-
lian, and I was pleased to discover our views were pretty well aligned.
Every individual, she wrote, "has an intellectual and spiritual obligation to decide for himself what he thinks and not to allow himself
to accept what comes from others without putting it through his own
reasoning process."

She added: "The important thing is neither your nationality nor
the religion you professed, but how your faith translated itself in your
life."

Her views were progressive for her time. Her only requirement
when it came to a person's faith was that "whatever their religious be-
lief may be, it must move them to live better in this world and to approach whatever the future holds with serenity."

Then I came upon something so shocking that I physically recoiled
from the book. In 1920, at the age of thirty six, Eleanor wrote to her
mother-in-law, "I'd rather be hung than seen at a gathering that was
mostly Jews." Even as late as 1939, she wrote to a former German class-
mate that "there may be a need for curtailing the ascendancy of the
Jewish people," but acknowledged that "it might have been done in a
more humane way by a ruler who had intelligence and decency."

When it came time for the afternoon prayer meeting, I was happy
to get away from Eleanor. Over the past nine months, she had become
something of a deity to me. Discovering such a fatal flaw caused within
me a crisis of faith. *Did my beloved equal rights advocate seriously dismiss
Hitler as a dunderheaded schoolyard bully?* I wondered during our prayer
service, which was led by Alice's husband, Jim, who'd just returned
from a business trip. He had a master's degree in divinity and a won-

derful booming voice that drowned me out and delivered me from humiliation during the hymns. I ate slowly at dinner and then practically plodded back to my cabin, wary of returning to Eleanor.

Instead, I decided to try mindful meditation, which is based on the Buddhist philosophy that much of our unhappiness comes from the need to hold on to things and react to them.

Life was full of things that provoked me—traffic jams and computer malfunctions, screaming babies on airplanes, annoying coworkers— and when faced with these things, I inevitably became upset. I formed a judgment about what was happening ("That baby is so annoying!" "I can't believe the subway is late again!"). Likewise, when I was worried about something, I became overly attached to that thought and treated the worry as something I must pay attention to until the problem was solved. Mindfulness would apparently train my mind not to react to these everyday stresses but to let them go.

During mindfulness practice, whenever thoughts came into my consciousness, I was to simply acknowledge the thought but not react or attach emotions to it. For example, if during my meditation a noise distracted me (a car horn), I was supposed to acknowledge the noise but then let it go. I'd simply go back to observing my thoughts. If a worry arose, instead of following the narrative of that worry by trying to solve it or predicting how it will affect my future, I'd simply notice the thought but not dwell on it. Mindfulness meditation teaches us to distance ourselves from our worries and stop engaging with them. Like a journalist who's covering an emotional story but remains an unbiased observer, instead of becoming emotionally wrapped up in it.

Eleanor, wise old dame that she was, knew the importance of a quiet mind. "I know many people who find it impossible to do anything unless they have complete calm around them," she wrote. "This must be because they have never learned to gain an inner calm, an oasis of peace within themselves."

The thought of Eleanor reminded me of the anti-Semitic remarks, and, shuddering, I put her out of my mind. I scooted back on my bed until my spine was straight against the headboard. I brought my attention to the rise and fall of my breath, as Dr. Bob had instructed. I observed it without trying to control it. After a few minutes, I found myself looking around the room. My eyes fell on the flat-screen TV. *I miss television. There's a new episode of Law & Order on tonight. Though I must say, I'm not enjoying the swaggering new assistant district attorney. And the medical examiner has done something ridiculous with her hair color.*

When I noticed my thoughts wandering, I brought my attention back to my breath. Inhale. Exhale. Inhale. *Wait, did I ask Jessica to feed my parakeets? And I hope Mom didn't forget that I was on a silent retreat this week so I can't call her back. She probably thinks I died in some fear-facing accident.* No, no, no, I scolded myself. *You're supposed to be meditating.* I focused on my breath for a minute or two, and then my heart started pounding. *I can't believe I still have so many fears to face! And time is running out! I'm going to run out of energy before I get to them all! Wait, what am I doing? This isn't working at all.*

It wasn't technology that had distracted me all this time, I realized, it was *me* that distracted me. My BlackBerry, computer, and TV were simply the devices I used to distract me from my worries—just as I'd been using sleeping pills to escape my worries at night. I decided to call it a day and try mindfulness again tomorrow.

My eyes drifted back over to the Eleanor book next to me on the bed. I opened it with a sigh. But, of course, Eleanor didn't disappoint. During the 1940s, her anti-Semitic beliefs faded away as she developed strong friendships with several Jews. Although she never publicly commented on the shift in her viewpoint, perhaps it's what she was thinking of twenty years later when she noted that "the narrower you make the circle of your friends, the narrower will be your experience of people and the narrower will your interests become. It is an important part of one's personal choices to decide to widen the circle of one's acquaintances whenever one can." Because of those friendships she became one of the biggest public supporters of Jewish causes. She lobbied Congress to ease immigration laws for Jewish refugees seeking asylum in the United States, and when she was unsuccessful, she gave the lawmakers a public spanking.

"What has happened to this country?" she scolded in her newspaper column. "If we study our history we find that we have always been ready to receive the unfortunates from other countries, and though this may seem a generous gesture on our part we have profited a thousand fold by what they have brought us." By 1947, she was calling for the creation of the Jewish state that would become Israel. "It isn't enough to talk about peace," she once wrote. "One must believe in it. It isn't enough to believe in it. One must work at it." I still felt betrayed by her anti-Semitic remarks, but in a way I respected her more now. It was one thing to try to change other people's minds about something, but she was willing to change her own mind first.

By my third day, I was going out of my damn mind. The rain made it impossible to go anywhere. Meals were the most exciting part of my day. Turning on my disco fireplace each night was a wildly anticipated event. The quote was "*Do* one thing every day that scares you," yet somehow *not doing* was even harder.

After the ceremonial flipping of the fireplace switch, I decided to try one of the tactics Dr. Bob had recommended to help me stop my fretting and focus on the here and now. "Start by compartmentalizing the worry in your life," he'd said. "Establish a thirty-minute 'worry time' during the afternoon where you write down everything that is troubling you. Every day that time—and that time only—will be solely devoted to worrying. Keep a pen and paper by your bed. If you start worrying while you're trying to go to sleep, write down the worry and leave it for your worry time the next day."

"But won't dwelling on my worries just make me more worried?" I'd asked him.

"On the contrary, your worries will seem more manageable. You'll realize that you don't have as many worries as you thought you did. It's not a hundred different worries; it's the same five over and over. After a while, they'll probably start to bore you. Once you're bored with something, you lose interest in it."

Another reason writing down worries is useful is because you can look back on them later and have proof of how unproductive worrying is.

"Studies have shown that when people are asked to write down their worries over a two-week period and predict what will happen, 85 percent do not come true," Dr. Bob had told me. "What does that tell you?"

"Most of the time there's nothing to be afraid of."

"Exactly." He looked triumphant.

Sitting cross-legged on the bed with a pad of paper, I started writing: "I'm worried my parachute isn't going to open when I go skydiving in a few weeks. I'm worried I'm going to bomb onstage doing stand-up comedy. I'm worried I'm going to get my ass handed to me by Mount Kilimanjaro. I'm worried that I still don't have a full-time job. I'm worried that I'll run out of money before I—" I stopped midstroke and the ink pooled on the paper where I'd left the tip of the pen. There were so many I's in that paragraph. Worrying, in addition to accomplishing nothing, is also self-indulgent, I realized. So often it's about you and your feelings. Yet another reason why it's good to limit it. Give yourself a bit of time, then get on with living.

I arrived at dinner on my last night to discover Margaret had gone home. Left without a word, obviously. I was surprised by how much I missed her presence. I knew nothing about her except she was married, according to her left hand, and favored nubby sweaters. Before bed, as I packed my books in my suitcase, it hit me that in lieu of gadgets, I'd been using the Eleanor books to distract myself too. Every time I started to face my worries this week, I'd turn to the books. They were just another form of avoidance. Wow, what a failure this week was.

After five days of storming, I heard the patter of rain stop almost immediately. I eased open my sliding glass door. I slipped into the backyard and walked to the labyrinth under the moonlight, taking care to step with intention. When I arrived, I told Alice I wanted to learn to just *be*. But *just being* required work. It wasn't enough to go on a silent retreat. You couldn't get rid of a few little objects and accomplish a major change in mental awareness in a matter of days. I'd have to work at it, not just for the rest of the year, but probably for the rest of my life. As I spiraled toward the center, I felt less guilt about the silent retreat not being an unqualified success. When walking a labyrinth, you could think you were going one direction, only to find out you were going the opposite way. Unlike a maze, there were no dead ends, just as there were no true dead ends in life—just opportunities to turn things around. Yes, I'd failed at meditating and been so distracted that I couldn't even focus on worrying, but we were all a work in progress. Even Eleanor, a onetime anti-Semite, became one of the biggest supporters of Jewish causes. "It isn't enough to talk about peace," she'd said. "One must work at it." She was referring to actual war, of course, but the same could be said of inner peace, as well.

And with that, I abandoned all thought. I continued walking the path, spiraling and spiraling, feeling the bricks cool and moist under my feet, making my way toward the center.

Chapter Eleven

~

The purpose of life is to live it, to taste experience
to the utmost, to reach out eagerly and without fear,
for newer and richer experience.

—ELEANOR ROOSEVELT

As a kid I often dreamed about falling. It was never clear where I was falling from, but it was obvious where I was headed. During these dreams my body would jerk violently, waking me up before I hit the ground. ("You're lucky," my childhood best friend had said in a tone of deep certainty. "If you hit the ground in the dream, you die in real life. It's a scientific fact.") For twenty-nine years, skydiving had literally been my worst nightmare.

I thought about all the physical risks Eleanor faced in her lifetime. In 1933, she took a two-and-a-half-mile ride underground deep into an Ohio coal mine. Coal mines were dangerous places, vulnerable to roof cave-ins, explosions, and flooding. When these disasters occurred, rescuing the miners was a difficult and often impossible task. But Eleanor wanted to witness the miners' working conditions for herself and ultimately deemed them "dark, dank, and utterly terrifying."

She later attended a meeting of the Southern Conference for Human Welfare in Birmingham, Alabama. Back in 1938, state law prohibited blacks and whites from sitting together at public gatherings. Eleanor strode into the racially divided auditorium and sat on the "black side" with her friend, civil rights leader Mary McLeod Bethune. When informed by the police that she was breaking the law and had to sit on the opposite side with the whites, she picked up her chair and placed it in the center aisle. She never stopped fighting for equal rights, even when threats were made on her life.

In 1958, she was about to fly to Tennessee to speak at a civil-rights workshop when she received a phone call from the FBI. "We can't guarantee your safety," they said. "The Klan's put a $25,000 bounty on your head. We can't protect you."

"I didn't ask for your protection," the former First Lady retorted. "I have a commitment. I'm going." At the Nashville airport she met up with a friend, a seventy-one-year-old white woman. They got into a car and drove off into the night alone, their only protection a loaded pistol placed on the front seat between them. If Eleanor could pack heat and face down a bunch of homicidal racists at the age of seventy-four, I could skydive. Though I knew my skydiving wouldn't change the world, I did think it could change me. And if I could take this kind of risk, maybe then, if I had a chance to make a difference in someone's life someday, I'd have the courage to do it.

Eleanor once said, "It is only by inducing others to go along that changes are accomplished and work is done." She was referring to leaders like Lincoln, Gandhi, and Churchill, who had to have a following in order to bring about real reform. I'd decided to co-opt this principle and use it as an excuse to make Bill, Chris, and Jessica jump out of the plane with me.

Matt was so terrified of heights that not only had he refused to skydive, but he couldn't even handle coming to watch us do it. He did, however, e-mail me an article titled "How to Survive a Skydiving Accident" the night before our jump. It was full of harrowing tales like the first-time skydiver whose tandem instructor suffered a heart attack and died in the middle of their jump, or veteran skydivers whose brains "locked" during the free fall, causing them to forget to pull the parachute cord.

"Oh, you're a riot. Thanks a lot," I wrote back. "Also, how absent-minded do you have to be to forget to pull your parachute cord? Who are they letting jump out of planes—Alzheimer's patients?"

"Good point," he replied. "I mean, what else have you got going on in that particular moment? Did you get the photo I attached, by the way?"

I scrolled down to the bottom of his e-mail and double-clicked the attachment. Suddenly my entire screen was filled with a photo of four stark naked skydivers beaming at the camera in midair. The picture was a disturbing testament to the bodily effects of falling at 120 miles per hour. The women's breasts were inverted to the point of resembling upside-down cereal bowls. I e-mailed it to my fellow jumpers with the subject line "Sorry, But You Need to See This."

Bill responded almost immediately from his BlackBerry. "Jesus, my eyes! Can we keep this a safe space please, Hancock?"

"Oh my God," Jessica added. "*What is going on with the boobs?*"

"That's what's going to happen to your boobs tomorrow, Jess," Chris wrote. "Sometimes, they stay that way."

"Seriously, you guys *cannot* make fun of me if I start crying up there," Jessica replied. "And we're doing tandem, right? People strapped to our backs?"

"I recommend doing everything in tandem," Bill wrote. "I've got a dude strapped to me at the drugstore right now."

"Yeah, it's always tandem for your first jump," said Chris, who'd skydived once before. "I think they do that because some people pass out on their first skydive."

"Really?" I said. "I didn't know blacking out was an option. Do I need to put in a request for that ahead of time or just tell them at check-in?"

The weird part was, even though this had been a lifelong fear, I wasn't as anxious as I'd thought I'd be. At the beginning of this project I'd gotten nervous about the mildest of challenges. Haggling with a vendor over a secondhand bureau, going to my first swing dance class. Even *thinking* about doing something intimidating had been enough to set off butterflies in my stomach.

But because the project was so big, it forced me to deal with my fear in a new way. As the year had progressed I'd noticed that the more I worried about future fears, the more overwhelmed I felt. I couldn't conquer one fear while worrying about climbing Mount Kilimanjaro and skydiving and whatever else I had coming up, or fear would consume my life. So to make the project more manageable, I took it day by day, focusing only on the challenge right in front of me.

That morning we found four seats facing one other on a train out of Penn Station. Jessica and Bill divided up the Saturday *New York Times*. Chris and I tried to arrange our long legs so that they weren't knocking against one other.

"The place is called Long Island Skydiving," I told the group. "Apparently they specialize in first-time jumpers."

"How long does skydiving take?" Jessica asked.

I cued up the company's website on my BlackBerry. "It says here 'Please plan to spend at least three hours with us on the day of your skydive.'"

"Three hours, or all of death's eternity," she sighed. "What's our stop again?"

"Speonk. It's about a two-hour ride."

"What kind of name is Speonk?" she asked.

"Actually, as I discovered this morning when I was looking up our directions, it was inspired by a Native American word meaning 'high place.'"

"It's also the sound your body makes hitting the ground when your chute doesn't open," Bill said, without looking up from his newspaper.

"Did Eleanor ever skydive?" Chris asked, changing the subject.

"The first commercial skydiving centers didn't open until she was in her seventies. But from what I know of Eleanor, given the chance, she totally would've skydived. She once took a mile-long toboggan ride at Lake Placid, and those things are no joke."

He searched my face. "You look pretty at ease for someone who's about to face one of their worst fears, by the way."

"I know, it's weird, right?" The last-minute freak-out had always been my specialty. As recently as a year ago I'd been at the front of a long line for an amusement park water slide when I'd suddenly decided I couldn't go through with it. All the kids behind me, along with their parents, had to make way for me as I trudged back down the stairs past them, trying not to meet anyone's gaze. So although I'd felt strangely normal all morning, I knew there was a good chance it just hadn't hit me yet.

"Just wait till you get on the plane. Longest fifteen minutes of your life." Chris grinned.

Two hours later, when the conductor announced that the next stop was Speonk, Jessica turned to me in a panic. "I need to have sex," she said.

"Excuse me?"

"You know, one last sexual encounter for the road in case I don't make it."

"Well, don't look at me."

"Or me," added Chris.

She looked over at Bill, who had folded the *Times* metro section into a hand puppet, which he was using to conduct conversations with nearby passengers. She turned back to us. "Actually, I think I'm okay."

The skydiving company consisted of a small landing strip next to a commune of trailers, one of which had a pirate flag perched on top. A sign out front read "Long Island Skydiving, Drop in here." Inside, one wall was covered with photos of happy customers in midflight.

"See, Noelle, look at how not scared those people are," said Chris.

"More importantly, look at how not dead they are," Bill added.

Chris leaned in toward one picture. "Oh my God. Is that Ricky Martin?"

"It is," drawled one of the employees, a sixtysomething man, who had come up behind us.

We all crowded in. The Latin pop star was frozen in the sky grinning at the camera, clouds lurking nearby like groupies. The wind was flaring his nostrils wide and pushing them up into two tiny parachutes. In a thick good ol' boy accent, the employee introduced himself as Cody. He contemplated Ricky for a few more seconds. "Some big calves on that feller."

We were ushered into a room full of folding chairs facing a TV monitor. There we watched a videotape in which an older man sitting behind a desk explained the risks of skydiving. The grainy production quality and the fact that the room was done entirely in wood paneling suggested it had been recorded in the 1970s, but the defining feature of the video was the man's beard, which was so long that it fell past the top of the desk. Jessica described the tape as "Professor Dumbledore tells us how we'll die" and took a picture of the beard with her digital camera.

"It looks like the kind of video a group of separatists in Montana sends to the president announcing that they're seceding from the Union," Bill said.

Soon it was time to sign the "If I die, no hard feelings" waivers. As an extra legal precaution, the company insisted that we say our names into a video camera and read the last paragraph of the contract out loud while they filmed us. With a perfectly straight face, Bill recited, "I, Bill Elizabeth Schulz, understand the risks that I'm about to undertake . . ." The general information form asked that we provide an emergency contact person, but cautioned, "Do not use anyone else who will be on the plane with you."

"Oh my God," Jessica whispered.

The four of us piled into a van that would drive us a few hundred yards down the airstrip to the plane. With the exception of the windows, every inch of floor, walls, and ceiling had been upholstered in gold shag carpeting. Cody swiveled around in the driver's seat. "Welcome to the Shaggin' Wagon," he said with pride as the engine coughed to life. "We bought this baby off some guy for $200."

"There's a lot of DNA on this carpet," Chris murmured.

Bill climbed in and admired the seventies decor. "But just think about how many mustaches have been in here!"

There were no seat belts—in fact, there were no seats. This didn't inspire a hell of a lot of confidence, especially when he drove us down the landing strip with the sliding door all the way open. When we reached the takeoff area, the instructors taught us the importance of arching our backs during free fall and picking our feet up upon landing to avoid being trampled by the tandem partner strapped to our backs.

The single engine plane had only enough room for two skydivers and their tandem partners. Dr. Bob told me that the longer you expose yourself to a fearful situation, the greater the reduction in anxiety in the future, so I wanted to go in the second round. Bill wanted to watch me freak out, so he decided to go with me. Chris and Jessica would go first. Jessica was paired up with a blond man named Timothy, who was slight with a gentle manner. He strapped her into a full body harness. Right before they jumped, Timothy would clip his harness into Jessica's, fusing them together for the skydive. He tested out the strength of the harness now, briefly clipping himself in behind her. Jessica was so petite that when Timothy stood up straight, her feet dangled off the ground. She looked like a child being carried in a Baby Bjorn.

"Okay, everything looks good," Timothy said, detaching the harnesses. "First group, we're up!"

"I love you guys," Jessica said shakily as she and Chris turned to follow their tandem partners to the plane. As the plane rolled away, there was an awful moment where Jessica looked at us apprehensively out the window. She suddenly seemed very young. For the first time I felt nervous, not for my own safety but for the safety of those I'd asked to come with me. Why had I brought all three of my closest friends? I remembered how your emergency contact couldn't be someone on your plane. I should've mixed in some lesser friends—spread the field a little. The sky was hazy and glaring; Bill and I squinted as we watched the plane circle higher over our heads.

"Is that them?" I asked when a speck emerged from the plane. Fifteen seconds later, a second dot followed. As the specks grew, I felt as though I was watching the process of gestation on fast-forward. Within seconds they'd evolved into wriggling human beings. After less than a minute the two chutes bloomed open, revealing their primary-colored insides to the dull, white sky. Chris's limbs were almost comically gangly next to Jessica's compact body.

"Like Peter Pan and Tinkerbell," I observed as they floated down. Even thirty feet up, I could detect a happy bewilderment radiating from their faces.

"You guys are so extreme right now!" Bill called out as they landed a few hundred yards away. "You'll never spell *extreme* with an *E* again. Just three *X*'s from now on."

"The first words out of my mouth in the air were, 'Holy crap!'" Jessica grinned. "I cursed through the entire free fall and spent most of the parachute ride apologizing to Timothy for cursing so much."

Chris was more wild eyed than I would have expected of someone who'd done this before. "There's something innately comforting about having another human being attached to you," he said. "But then that person betrays you by hurling themselves out of the plane while you're still attached to them."

"How did this compare to the first time?" I asked.

"This skydive was more"—he paused, clearly searching for a safe adjective—"*extreme* than my first one. When I was screaming and plummeting toward the earth, I bet my tandem partner was thinking, 'Wait, I thought the other guy had the girl attached to him.' "

I inherited Jessica's instructor, Timothy, as my tandem partner. Bill was teamed up with an instructor named Sebastian, who is what you'd get if the Marlboro Man mated with the Old Spice guy and decided to raise their son in the UK. He clapped his hand so hard on Bill's slim shoulder that he stumbled forward a bit.

I leaned over to Bill and whispered, "There's no way you're getting through this day without receiving a fist bump."

"There is an art or, rather, a knack to flying," Sebastian said in his rugged British accent. "The knack lies in learning how to throw yourself at the ground, and miss." He grinned naughtily.

Sebastian was six feet six and dazzlingly handsome, with twelve thousand jumps under his belt to Timothy's three thousand, but I was glad he wasn't my partner. As the four of us walked to the plane, I overheard Sebastian telling Bill, "It's not the dying I fear if me chute don't open—it's the livin'. Lyin' there on the ground, staring at my intestines, knowing I'm gonna spend the rest of me life in a wheelchair, communicating via the one eyebrow that still moves . . ."

When we reached the plane, Timothy pointed to a rectangular piece of metal just above the plane's wheel. It was a few inches wide and stuck out about two feet. "See this step? When it's our turn to jump, I'm going to first have you step out of the plane and onto that plank, but don't look down. When people look down, they tend to freak out. Look at the propeller instead."

The propeller? I thought, hoisting myself inside the plane. This was the best alternative focal point they could come up with? ("If you find yourself overwhelmed by the sight of the ground ten thousand feet down, set your mind at ease by focusing on the cluster of revolving blades a few feet away . . .")

Again, no seats, just some flaccid safety belts attached to the floor. I scrunched into a ball in my designated spot with my back to the pilot's chair. Timothy was curled up on the floor directly in front of me, so close that our shins pressed together. I was shoulder to shoulder with Sebastian, who was on my left, while Bill squatted on the floor next to the pilot's seat where the copilot's seat would normally be. He was cautioned not to touch any switches on the dashboard or risk turning us into a newspaper headline. Looking out the window to my right, I could see Chris and Jessica waving as we rose from the runway and teetered off into the sky.

I kept waiting for myself to panic. To start crying and asking to turn back. But I was shockingly composed sitting there in this tiny plane—my third tiny plane this year—knowing that I'd soon be jumping out of it. Being afraid seemed almost . . . pointless. I remembered Eleanor recounting how, when Franklin was Assistant Secretary of the Navy, he had her tour insane asylums and report back to him on the conditions.

" 'I cannot do this,' I thought. I was terrified of insanity," she wrote. "Then I realized that I was the Assistant Secretary's wife. This was my job. I had to do it whether I could do it or not." Fear was my job now. It was a place I had to go to every day, so why bother whining and resisting it?

"We're at five thousand feet now," Timothy shouted over the engine. "Shouldn't be long."

The day was windy, and the aircraft pitched boozily as the pilot struggled to keep us level. With the exception of the Shaggin' Wagon, the plane seemed like the least safe part of this operation. During one especially lively dip, I involuntarily grabbed Sebastian's knee. When we careened sharply to the right a minute later, I wrapped my arm around his entire leg.

"I bloody love this girl!" Sebastian crowed and everyone laughed.

I was actually eager to skydive now. Just get me off this plane. I reminded myself that in ten minutes this would all be over and I'd be back on the ground with Chris and Jessica.

"We're at eight thousand feet," Timothy updated me.

This was my cue to turn around on my knees so he could position himself behind me and clip us together. Timothy pulled the straps tight. In the front, Bill was making his way onto all fours, trying desperately not to disturb any of the instruments on the dashboard in front of him. As Sebastian leaned over Bill to clip the two of them together, he pretended to vigorously sodomize him.

"Yeah, yeah, *yeah!*" Sebastian shouted with each mock thrust. "You like that, mate?!"

"About thirty seconds now," Timothy said. He'd crouched behind me on the balls of his feet.

Just breathe, I told myself, remembering what Dr. Bob had once told me about safety behaviors. "When people are afraid, they hold their breath," he'd said. "They're trying to close themselves off from the fear, but trying to rid oneself of fear never works. I want you to breathe into the fear. Immerse yourself in it. As you inhale, imagine yourself taking all of that fear in."

"We're at ten thousand feet!" Timothy chirped. "It's about that time!"

Bill and Sebastian would go first because they were closer to the hatch. Sebastian kicked open the door. Freezing wind rushed into the plane. The two of them positioned themselves in front of the gaping hole where the door had been. Clumps of my hair enthusiastically leaped into the air, as if waving farewell. Should I watch them jump? I didn't know if it was a good idea. I turned away, but watched out of the corner of my eye. Unable to resist, I turned back just in time to see Bill and Sebastian lean over and get sucked out of the plane at an alarming speed.

"Now it's us! Go! Go! Go!" Timothy urged. I inched toward the open door on my knees, pulling him along.

"Okay, Noelle, put your foot on the step!" he yelled.

I placed my foot on the metal board, staring hard at my shoe, blocking out everything else. There was something very soothing about seeing this sneaker, which was so familiar to me, while ten thousand feet above the ground.

"Good!" Timothy shouted. "Now stick your head out! I'm going to count to three!"

Still looking at the sneaker, I plunged my head into the ninety-mile-per-hour wind. During ground training, Timothy explained that during the countdown we were going to rock back and forth twice, then roll out of the plane on the number three.

"One!" We leaned out.

"Two!" We leaned back toward the plane.

"Three!" Timothy rolled us into the sky and then we were dropping headfirst at two hundred feet per second.

In the first few moments out of the plane, I had two thoughts. Every skydiver I'd talked to had told me, "There's no stomach drop like on a roller coaster. You feel like you're floating, not falling." So my first thought was, *They lied.* There was a stomach drop feeling. It was only for a second or two, but still. It was worth mentioning. I felt a flash of irritation at Chris and Jessica for not warning me. And while I didn't feel like I was dropping two hundred feet per second, "floating" was something of an understatement. I was very aware my body was hurtling toward the earth. My second thought was, *My God, this is high. I can't believe that I'm going to be twice this high at the peak of Kilimanjaro.*

Then there was no thought. All of my senses were overstimulated. The sky and ground were twisting before me. I had no idea where Bill was. The sound was almost deafening. I was ripping through the heavens. The wind was pushing my cheeks and lips into an idiotic grin, which was pretty much how I felt. Timothy rotated us around to give me the 360-degree view. It had been overcast before, but now the sky was clear and the sun flashed between the clouds. I could see Fire Island and the bay twinkling in the distance. The horizon was a glowing circle around me that cast a pale, almost ethereal light over the earth. There was none of the harsh forest greens and deep ocean blue I was used to seeing from airplane windows. There was no geometric patchwork dividing the ground into sections. It was as if the world had been repainted using pastels. Everything blended together harmoniously.

"Wow," I gasped over and over. My windblown lips could barely form the words. "It's. So. Beau. Ti. Ful."

When Timothy tapped me to signal he was about to pull the chute, I couldn't believe forty-five seconds had already gone by, it had felt like five. Timothy pulled the cord. I'd been worried it would open with a painful jerk, but it was more of a gentle upward tug. For a few seconds, the roaring wind disappeared and I was surrounded by the most absolute silence I had ever known. The quiet was profound. Once again I was gasping in awe. Then the sounds of the parachute flapping in the breeze kicked back in and the moment was gone.

"What do you think of the view now?" Timothy asked. I'd almost forgotten about him back there.

I looked down. The ground still had that dreamy quality. It looked unrealistic. Yet my dangling legs looked too real. My black spandex pants, the Nikes—they were too much in focus, the colors too vibrant. They seemed ridiculously out of place against the soft-hued earth. I was reminded of the "outdoor" scenes in 1960s movies where they shot actors in front of prefilmed background footage, then turned on a fan to simulate wind.

"I did it!" I squealed.

"Here, hold these," Timothy said, giving me the steering handles attached to the ends of the parachute. "I'm going to unclip your waist harness, which should loosen up your wedgie."

Before I could respond, Timothy had unlatched my waist and I dropped about six inches. For a few startling moments, I thought I'd been cut loose. Then the shoulder straps caught me under the armpits. I looked up at the handles, which I was gripping with all my strength. Everyone had told me that the parachute ride was the nice part, but now I looked up at the handles and wondered, *What happens if I accidentally let go of them? Will we go plummeting to the ground?*

"Now pull down with your left hand and steer us to the left," Timothy instructed.

"To be perfectly honest, I'd rather you drive."

He took hold of the toggles. "You can let go."

"Do you have them?" I asked.

"Got 'em."

"You're sure?"

"I'm sure."

Still, I looked up to double-check that he had them before taking my hands off. A few hundred yards away, Bill and his parachute were completely horizontal as Sebastian took them on some wild corkscrew turns. They were dropping fast, too fast.

"Oh no!" I cried. "Are they in trouble?"

Timothy chuckled. "No, that's just Sebastian being a daredevil."

I sighed with relief. "Okay, well, please don't do that to me." Instead, we simply hung for the next four blissful minutes.

The airstrip appeared below us. I could see Chris, Jessica, and Bill—who had only jumped twenty seconds ahead of me, but who had reached the ground a full five minutes before me because of those corkscrew turns—bobbing up and down in celebration.

"Because it's a windy day we're going to come in pretty fast," Timothy said. "Depending on how we're positioned as we land, at the last second I'm either going to tell you to sit or stand."

Dude, we *were* coming in fast. Like, *really* fast. I felt as though I was about to jump from a moving car. When we were level with the trees, I asked nervously, "Sit or stand?" No answer.

"Sit or stand?!" I shrieked. The ground was a few feet away.

"Sit!" Timothy commanded. I held up my legs while Timothy's sneakers scrambled for purchase and eventually clomped to a halt.

"You did it!" Jessica squealed running over with Chris.

I was too bewildered to say anything. What came out of my mouth was a cross between a laugh and a horse whinny.

Bill was already deharnessed and drinking a soda. "Nice to have you back, Hancock," he said, giving me a significant look so I knew he was referring not to our skydive but to our heated exchange at the end of the shark cage dive when he told me that I'd changed.

"Your corkscrew turns were amazing!" I told him.

"I was completely terrified," he admitted, but quietly enough so only I could hear. "But I was afraid that if I said something, he'd just go faster."

They gave us our "diplomas," and we took a group picture in front of the skydiving sign. Afterward Bill clapped his hands together and said, "You guys want to do a nude one really quick?"

A half hour later we were at the train station. It was an outdoor platform, but we were sitting on a bench inside a covered waiting area. A giddy energy hung in the air, as if held in by the Plexiglas walls. Bill looked us over approvingly. "From where I'm sitting, I'm looking at three totally radical individuals right now. Look, we're so radical that lady won't even come in here," he said, pointing to a woman talking on her cell on the platform outside. I pulled out my skydiving certificate. It read:

LONG ISLAND SKYDIVING CENTER
HEREBY AWARDS THIS CERTIFICATE OF TANDEM FREEFALL SKYDIVE TO

Noelle Hancock

WHO, ON THE 9TH OF MAY IN THE YEAR 2009, DID EMBARK ON A MOST FANTASTIC JOURNEY. EXITING HIGH ABOVE THE GROUND FROM AN AIRPLANE IN FLIGHT, CASTING FATE TO THE WIND AND FALLING FREE. MAY YOU ALWAYS ENJOY BLUE SKIES ABOVE, AND MAY YOUR LANDINGS BE FOREVER SOFT.

"It sounds like it was written by an extra from *Bill and Ted's Excellent Adventure*," Bill grumbled, "which isn't that surprising considering the strain of dude we had instructing us." He still looked slightly shaken from his wild descent with Sebastian.

"The instructors definitely wrote that while smoking weed in that trailer with the pirate flag," Chris agreed.

"I actually think it's kind of sweet," Jessica said.

We turned to her in surprise and she blushed. "I mean, when I was floating down I was thinking all of these hippy-dippy 'I love the universe and everyone in it' thoughts. And it *was* a totally freeing experience." She added, with a defiant little nod, "So, yeah, I get whatever the hell they're talking about."

"I brought a surprise," Chris said. Grinning slyly, he pulled a bottle of rum from his backpack. "Shall we celebrate?" We passed it down the line, each us taking a long pull straight from the bottle. Then we did it again.

I was so utterly content that I was slightly wistful. It was the feeling I got when I was about to finish a really great book. I was nostalgic for this moment even as I was still in it. Soon the train would come and whisk us back to the city. I wished I could sit there forever in this train station in the middle of nowhere with the three people in the world who would jump out of a plane for me.

Chapter Twelve

⌒

A mature person is one who does not think only
in absolutes, who is able to be objective even
when deeply stirred emotionally, who has learned
that there is both good and bad in all people and
in all things, and who walks humbly and deals
charitably with the circumstances of life.

—ELEANOR ROOSEVELT

"It was the weirdest thing," I told Dr. Bob when he asked how
skydiving went. "When I was about to jump out of the plane, I
wasn't afraid at all." I brushed a clump of my hair off my forehead, remembering the loose strands had danced in the air when they
flung open the door and the wind rushed in; yet somehow I'd stayed
utterly composed. "I told myself, 'You don't have to get through this
long scary ordeal, you just have to get through the present moment.'
When I thought of skydiving as just a collection of moments, I realized there were maybe three seconds of scariness—the part where I
was stepping across thin air to put my foot on a ledge outside the plane.
And when that started to feel scary, I brought it down to an even more
micro level by concentrating on my shoe. I was in this mental place
that was supercontrolled, yet free."

I noticed Dr. Bob was smiling at me knowingly. "What? Why are you looking at me like that?" I asked.

"You were practicing mindfulness."

"I was?" I said wondrously. "I was!" I'd tried to practice mindful meditation for months, but up in the air it just clicked.

Now that I'd figured out how to conquer physical fears, it was time to start facing more emotional ones, like my fears about my relationship with Matt, which I'd shoved into the basement of my mind since the Nantucket trip. So much of the time in relationships we're peering into the future, trying to predict what may lie ahead. Maybe the real answers were in the past. And if I wanted to avoid screwing up my relationship with him, I needed to examine the mistakes I made in past relationships.

"It is easy for us to be quite misled about ourselves, about our bad qualities as well as our good," Eleanor once wrote. "And it is impossible to proceed with the right motives instead of the wrong ones as long as we have any serious misconceptions about ourselves."

There were two types of people in this world: people who stayed friends with their exes and people who didn't. I fell squarely into the latter category. I hadn't spoken to my exes in years, which made them a pure, untapped resource. Ask your friends or family members to talk about your flaws and they'll soft-pedal it. But your exes will give it to you straight. You're no longer friends so they don't worry as much about protecting your feelings. Also, significant others are privy to your dark side in a way that family and friends rarely are. It's an unvarnished opinion. As Dr. Bob pointed out, perfectionists organize their lives around avoiding mistakes. The thought of revisiting my past failures and rejections made me feel more vulnerable than any physical challenge I'd undertaken. But it would be worth it if it would help me avoid making the same mistakes with Matt.

Frankly, I was shocked that my two college boyfriends, Isaiah and Ben, agreed to meet with me. There was no upside to the interview for them, after all. Both relationships had ended badly.

Isaiah and I had met during my senior year of college. He was the captain of the basketball team and we dated for ten months. In the beginning he was affectionate and caring; then over the course of the year he grew distant. I asked him what was wrong and he wouldn't tell me. Toward the end of our relationship he positively shut down. He rarely wanted to hook up with me. The more distant he became, the harder I clung. My presence seemed to make him irritable. I no longer made him laugh. He'd been really taken with me in the beginning, so I knew it must have been something I'd done. Trying to get us back to the place we'd been before, I threw myself deeper into the relationship, which only made him retreat further. One night while sitting on a beach together, I said: "I love you." When he said nothing back, it was a symbolic moment: I'd been having a one-sided relationship with myself. A few months later he dumped me—horrendously—on my birthday, via cell phone, from another girl's party. It was one of the more distressing periods of my life, so it was difficult to reach out to him. Yet he sounded happy to hear from me and, to my surprise, had incredibly positive memories of our relationship.

"But you grew to resent me," I said. "Why? Was it something I did?"

"It had nothing to do with you or us. I was in a depressive state. I was nearing the end of college, I'd been playing poorly all season, and I realized that basketball, the thing I'd lived for my entire life, was over for me. That was the most difficult year of my life, but it would've been much more difficult without you. You were the happier part of my day. I felt safe when I was with you."

I was dumbfounded. *"Seriously?"* For so many years I'd wondered how I'd screwed things up and it turned out that it hadn't been me at all.

With Ben I'd had the opposite problem: I hadn't loved him enough, yet I hadn't been willing to let him go either. After a year and a half together, I semi–broke up with him; then we sort of got back together while seeing other people at the same time (note: this *always* works out well). Our neither-here-nor-there relationship carried on for another year, but it had curdled. What was once sweet turned sour, chunks of bitterness clogging all of our interactions.

"What mistakes did I make in our relationship?" I asked him. We were having lunch in New Haven, where he now worked as a building manager for the Yale Divinity School.

"I wish you'd been more firm with me and hadn't allowed us to keep hooking up," he said. "I wish I'd had enough self-respect to stop trying to get you back."

"Why do you think it got so bad?"

"For that last year and a half we didn't have the security of a relationship, so we were jealous and insecure. And we were immature."

Immature. That was one word for it. We behaved in ways we never had before and haven't since with anyone else. Ben left drunk messages at five in the morning about how much he loved me and hated me. Once, after he copped to fooling around with one of my friends, I stormed into his closet and pulled from the hangers every nice shirt and sweater I'd ever bought him.

"I'll be damned if you're going to go around picking up other chicks using my good taste!" I shouted over the pile in my arms.

"You can't do that!" he protested. "Those were presents!"

"And *presently,* I'm taking them back," I said, marching off. Begrudgingly, I returned the repossessed shirts the next morning after a firm lecture from my roommate, Amanda, on the spirit of giving.

"Okay, last question," I said at the end of our interview. "People change a lot in relationships. Or sometimes the person that they always were just becomes clearer. How did your opinion of me change from the beginning of the relationship to the end?"

He thought for a moment, started to speak, then stopped. I nodded encouragingly and he admitted with a grimace, "By the end of our relationship, you were no longer attractive to me. Like, I knew I'd once thought you were pretty, but by the end you'd become ugly to me." Well, that stung, but I could tell that it had pained him to say it. To break the tension, I laughed—then so did Ben. I thanked him for his honesty.

The third ex-boyfriend, Josh, was my high school sweetheart and most serious relationship of the three, and therefore the scariest. My fingers were actually shaking as I typed my pitch: "Hey, I'm working on this project where I try to face all my fears before age thirty so I'm going back and interviewing all my ex-boyfriends about our relationship (because how scary is that—right?! LOL!). Would you want to sit down and chat sometime soon? I can come to D.C."

"Hey there. Sounds dangerous! Ha!" Josh wrote back. "Can't wait to hear more about it. You should come down the weekend after next! You can stay on my couch."

My surprise that he agreed quickly turned to worry. I hadn't mentioned Matt in the e-mail and didn't know if Josh was dating anyone himself. What if he thought this was an elaborate booty call? Surely he didn't think I needed to cross state lines to get some action? But a few days later another e-mail from Josh arrived.

"By the way," he wrote, "my girlfriend, Monique, pretty much sleeps at my apartment every night. But don't worry—she is happy to have you stay with us. We're really looking forward to it."

Both relief and alarm surged through me. Relief because I now had a non-awkward reason to write back, "So am I! And the next time you guys are in New York you'll have to meet my boyfriend, Matt." The uneasiness came from the realization that not only would I be hanging with my ex-boyfriend for two days, but I'd also be sleeping fifteen feet away from him and his current girlfriend. Suddenly my decision to stay the night felt a little aggressive.

"I can't believe that you're doing this," Jess said when I called her in a mild panic the night before I left for D.C.

I opened my laptop. "What? Going to D.C. to spend a weekend with my high school sweetheart and his practically live-in girlfriend? Or planning to cyberstalk the girlfriend before I go?"

"Both. You know what? I'm coming over," she told me. "I'm just leaving the gym, so I've showered but will be adorably sans makeup. Adorably meaning frighteningly."

"Bring a bottle of wine." I was too addled to even remark on the revelation that Jessica had joined a gym.

"I brought wine and Chris," Jessica said when I opened the door to find the two of them on my doorstep. As soon as we had poured the wine, Jess hauled my computer onto her lap.

"First of all, let's get a visual and see what we're working with." Within seconds, she'd pulled up Josh's Facebook page and had identified the girlfriend via one of his photo albums.

"Oh shit," I breathed, reaching across Jessica and pressing a button to enlarge one of the pictures. Lustrous black hair and rich olive skin filled the screen. She was stunning. I made a despairing face at Chris and Jessica.

"Okay, so she's uncomfortably pretty," she conceded.

"She looks really fun, too," I said miserably.

"You can tell from a photograph that she's fun?" Chris asked dubiously.

"It's her earrings—they're fabulous." Suddenly a horrifying thought entered my mind. "Oh God, do you think they'll have sex while I'm there? Is that a turn-on? Like having sex when your parents are home?" I was down to half a pill a night, but it still took me at least an hour to fall asleep.

"Hell yeah!" Jessica said. "A girl's gotta mark her territory when her man's ex is in the next room."

"We would," Chris agreed.

"Yeah, but we're petty and small." I buried my face in a pillow. "You guys! If I hear them having sex, I will seriously go into cardiac arrest."

"Earplugs," Jessica advised, raising her wineglass and toasting their existence.

"What if they're all lovey-dovey in front of me and it's weird?"

"Why does it matter?" Chris asked. "You've got a perfectly wonderful boyfriend of your own to kiss and hump."

"I know, you're absolutely right." He was so wonderful, in fact, that when I'd asked him if he was okay with me visiting Josh and his girlfriend for the weekend, he'd replied, "This is someone you dated when you were kids and we're not kids anymore. So, aside from the fact that I trust you unhesitatingly, it never really occurred to me to worry." He had no reason to worry. It wasn't that I thought I'd have feelings for Josh, but that I'd have feelings for Monique—jealous feelings. It was one of my worst qualities. I had the capacity to be jealous over guys I'd dated years ago, guys I didn't even like. It was incredibly childish and surely spoke to some larger insecurity that should be explored with Dr. Bob, but first I had to get through this weekend.

Josh had one of those boisterous voices that grabbed you by the shirt-front and said, "Now hear this!" He was a native Texan yet seemed like a scrappy kid from 1930s Brooklyn. One time when he was walking up the stairs, I saw him turn to the popular cheerleader directly behind him and say, "I'm gonna let you look at my ass, okay? It's sensational, but try to contain yourself." He was brash and inappropriate, and I loved him immediately. I figured out his schedule and rearranged my routes so that I ran into him between classes. I flirted with him stridently, slapping his butt as I passed him in the hallway.

"Hey, no touching the merchandise!" he'd cry.

We'd been dating for a few weeks when he invited me to the ROTC formal. I borrowed a friend's blue satin dress that implied, errone-ously, that I had breasts, and my mom blow-dried my hair for the occasion. We were in the middle of a slow dance when a dozen of Josh's ROTC buddies approached him eagerly, saying, "It's almost time!"

"Time for what?" I asked.

"Actually that's something I've been meaning to ask you about."

"Okay . . ." My heart raced. He was going to ask me to be his steady girlfriend. Here in front of everyone!

"Since my first ROTC ball freshman year, I've had this ritual of getting down on all fours and galloping around the dance floor while bucking like a mule," he explained. "But I was thinking this year maybe you could ride me around the dance floor instead."

By the time I realized he was serious, the partygoers had formed a massive circle around us. Someone even had a video camera. Josh got down on his hands and knees and looked at me over his shoulder with a "how 'bout it?" expression. When your crush requests you take part in an elaborate donkey fantasy before your contemporaries, there's really only one way to react. So I hitched up my floor-length gown, climbed on his back, and held on as tight as I could.

The relationship lasted a year and a half. In that time, we engaged in romantic one-uppery with the kind of vigor that doesn't make it past your teen years. He hijacked the school's PA system and asked me to prom over the loudspeaker. On his birthday, I sent him a singing telegram. I took out an ad professing my adoration in the *Houston Chronicle*. He covered my car in roses while I was at work.

"That could never happen in New York, by the way," Jessica interjected. "Someone would steal that shit."

"And the hood ornament, just to make a point," Chris added.

The night he first said "I love you," Josh showed up on my doorstep in a three-piece suit and took me to a candlelit restaurant in downtown Houston. Strolling hand in hand through a nearby park afterward, we came upon a majestic fountain. He scooped me up, carried me into the fountain, and slow-danced with me in his arms. "I love you," he said. Then, grinning wickedly, he dunked me, completely soaking me and my cocktail dress. I shoved him down and he splashed me. Eventually, our peals of laughter drew a crowd, and everyone applauded as we dragged ourselves out and bowed.

My relationship with Matt was steady to the point of being predictable. Josh had written long letters, detailing every aspect he loved about me and how he would die for me. Matt gave me a card on our third anniversary that read, "I'm grateful to have you in my life. I love you. Love, Matt." *I'm grateful to have you in my life.* It was lovely, but it was also something I said to my friends. Hell, I think I'd even said it to my hairdresser. Was that enough passion? Would that sustain me over a lifetime?

Jessica smiled thoughtfully, swirling her wine around the glass. "In New York, we can have the best of everything. It's a city with limitless options. So we get accustomed to thinking that there's always something better out there, because there usually is: a better apartment, a better job, a better meal at a better restaurant around the corner. We're never satisfied. This city trains us to worry about the possibility of something better, so we're unable to recognize when we actually have The One. Why do you think New Yorkers get married later than the rest of the country?"

"Why did you and Josh break up?" Chris asked.

"He was a year older than me and he went to college in Boston. We stayed together long distance, but it was too hard being in such different worlds," I said. "You know how I knew it was over? On our eighteen-month anniversary, I'd planned an elaborate scavenger hunt for him, leaving clues at various landmarks in our relationship, leading to his anniversary gift, which I left at the fountain where he told me he loved me. But when I'd handed him the first clue, he'd sighed and looked annoyed. 'How long is this going to take?' he asked. 'My mom needs the car.' "

The next day I stepped off the bus in D.C., uncomfortably aware that Josh would see me before I saw him. Finally I caught sight of him waving happily behind the tinted window of his silver station wagon. His hairline had receded slightly, but otherwise he hadn't changed at all.

"How about hamburgers for lunch?" he asked when I climbed in.

Between bites, we caught up on each other's families. I was happy to hear his mother's breast cancer was still in remission. He was a little appalled to hear that my little sister, who had been an infant when he and I dated, was now fourteen and had a boyfriend of her own. He told me he had recently applied to several business schools and was waiting to hear back.

After lunch he drove me by the White House and various monuments whose significance was pointed out through the car window at fifty miles an hour.

At a red light, he turned to me and asked, "So where do you want to go to do the interview? At a coffee shop somewhere?"

I hesitated, suddenly overtaken by a wave of vulnerability. I was worried that if I were looking into his eyes and he told me something I didn't want to hear, I might cry. And I didn't want my blubbering to alter his answers. But I sensed I could keep it together as long as we were sitting side by side, each of us staring straight ahead.

"How about we just drive around and I ask you the questions in the car?"

If he thought this was a bizarre request, he didn't show it. "All right. But you're buying gas later."

He maneuvered the station wagon into Hains Point in East Potomac Park. We were driving in a big circle, I realized. There was a soothing, meditative quality to these laps. Psychoanalyst Carl Jung believed that the act of drawing circles encouraged one to venture into his or her subconscious. The first objects that children draw are circles. To Jung, they represented the struggle and reconciliation of opposites and the eventual reunification of self.

Including Matt, I'd dated only three guys since Josh. The few times I'd had coffee with Josh in the last ten years, I'd never once asked about who he was dating. I took a deep breath.

"Okay, first question. How many girlfriends have you had since me?"

He laughed in an affectionate way that made my question feel very young. "Gosh, let me think for a second. That's hard to quantify. Does casual dating count or just relationships over a year?"

Josh, I was both surprised and not surprised to learn, had a substantial inventory of exes. There was an entire category, in fact, devoted to girls named Amy. With sudden embarrassment, I realized that our relationship wasn't one of the defining love affairs of his life, that I hadn't meant to Josh what he'd meant to me. I was just a blip, a non-Amy.

"How did you and Monique meet?"

"Mo-Mo and I have been friends for years, but we realized last year our chemistry was more than friendship. It's been absolutely wonderful."

This last line stung a little, as did his use of "Mo-Mo." I busied myself with my list of questions and tried to betray nothing. Now it was time to dive into our relationship.

"What attracted you to me initially and what ultimately turned you off?"

"Your self-confidence, your vivaciousness. You could go into any room and be part of it immediately." He said with a smile, "You could tell a story better than anyone."

This saddened me. The last time I'd felt like part of a room, I'd smoked some bad pot and believed I was a piece of furniture. I could hold my own one-on-one, but as I'd gotten older, groups had started making me nervous. "Why do you always clam up at dinner parties?" Matt used to ask before I stopped going to dinner parties. Everyone else seemed to have funnier, more intelligent things to say, and the more people who spoke, the lamer my opinions felt. The more time that went by without my saying anything, the more significance was attached when I finally did say something. ("She waited all this time to say that?" I imagined them thinking.) When I finally did speak, I'd be okay for a sentence or two, then I'd start to panic, lose track of what I was saying, and abruptly wrap things up by concluding, "So . . . *yeah*," baffling everyone in the room. It started in college. I began writing out a few talking points the night before and kept the piece of paper on my lap during class in case I blanked out. I made sure to recite my talking points toward the beginning of class, before they could be judged in comparison to what everyone else had said. I contributed just enough to get my participation grade and no more.

"And what about me didn't you like?" I was scared that he'd rattle off a laundry list of negative characteristics I didn't even know I possessed. Things that I couldn't change. I tried to remember what Eleanor said: "A mature person is one . . . who is able to be objective even when deeply stirred emotionally."

"You were too possessive," he said immediately.

One incident came to mind. Josh and I had been in line outside an eighteen-and-up club and a limousine had rolled up bursting with a bachelorette party. Standing up in the sunroof had been a group of drunk women with big hair that brought to mind tulips in a too-short vase as they wavered back and forth.

"Lookin' good, ladies!" Josh had shouted. No sooner had they beckoned him with their long acrylic fingernails than he'd started running toward the limo, leaping onto the roof and diving headfirst into the sunroof. His legs were still hanging out. I could see groping hands full of fake fingernails reaching up from inside the car, trying to pull him all the way in. I'd marched over, plunged my hand into the sunroof, and dragged him out by the back of his pants.

"Well, to be fair, you did *cheat* on me. It's not like I didn't have reason to be paranoid." I could laugh now, though his infidelity had been devastating at the time. "Do you have any regrets about our relationship?"

"I really regret cheating on you." A few months after we started dating, Josh and his friends had taken a trip to Austin, where he hooked up with a UT sorority girl. "Do you regret punching me in the face when you found out?"

"Not in the slightest," I said cheerfully. "Have you ever cheated on anyone else?"

"No. How could I after seeing what that did to you? We never got past it."

He was right. The cheating incident upset the power balance in our relationship. For the rest of the time we dated, it came up every time we disagreed about anything, no matter how small. Sometimes while we were kissing, I'd picture him making out with the sorority girl (who I knew—after demanding every sordid detail—was short, curvy, and brunette, my polar opposite) and my mood would instantly sour. It haunted me for years, long after we'd broken up.

"I'm a big flirt, though," Josh said. "A girl can't hold on to me too tight."

"What's Mo-Mo like?" I asked, testing out the nickname.

"Mo-Mo is very chill and not possessive. She's the most easygoing person I've ever met."

"Okay, anything else you didn't like about me?"

"You hated traveling, and I remember thinking I could never be with someone who hated traveling."

It was true. Even on vacations that required nothing more than drinking on a beach, traveling left me uneasy. I hated the feeling of being in transition. There was a nervousness in my stomach like when I was having a problem but hadn't yet figured out the answer. The problem only felt solved once I'd returned home, my belongings were unpacked, and I was back to my routine. Josh spent the year after college backpacking his way across three continents. He ran with the bulls in Pamplona, hiked the jungles of Thailand, climbed glaciers in Argentina, and found work picking grapes on a vineyard in France. International travel—especially alone—had been on my list of fears from the start. I was tempted to tell Josh about my plans to climb Kilimanjaro to show him how much I'd changed. I knew he'd be both jealous and impressed. But that wasn't what I'd come here to do, so I stuck to my questions.

"Why do you think we didn't work out?" I asked.

He thought about this for a few seconds. I thought he'd say something like "We were too young" or "After I went away to college, the distance became too much of a strain." Instead, he said, "I was more willing to throw myself out there and you liked things you were more used to." I blanched. One of the things I'd loved about Josh was how he'd forced me out of my comfort zone. But maybe to him I'd simply been dragging him down? The idea hit home because it was one of the things that worried me about my relationship with Matt. Every time he criticized me for not being more social, I wondered how long until he just got fed up and left me for some girl who loved dinner parties and didn't need a therapist to coax her into trying new things.

Josh pulled over near a thick assemblage of bushes and opened the car door. "After an hour of driving around, my bladder is at breaking point," he said. Then, after making sure a cop couldn't see him, he headed toward the bushes and peed.

The plan was to go back to Josh's place, meet up with Monique, and head out to a bar. That way if meeting her was awkward, we could drink after. Immediately after. Josh and I were sitting on opposite sofas chatting in his living room when a key turned in the lock. My breath caught. This meeting, I knew, would set the tone for the rest of the weekend.

"Anybody home?" a pleasant female voice called out.

"We're in here!" Josh called.

When Monique walked in, I saw it was worse than I'd thought. She was even more beautiful in person. The Facebook photo had also failed to convey her perfect sumptuous breasts.

She shook my hand. "I'm so happy to finally meet you," she said so warmly I actually believed her.

Josh scooted over and made room for her on the couch. I was relieved that they didn't kiss hello. We chose a safe subject and chatted about work. She was an international program specialist at the U.S. Department of Commerce. She asked me questions about my Year of Fear, and I recounted some of my latest adventures. At one point she reached over and stroked his hair. It was hair that I'd stroked ten years earlier and would never stroke again, hair that had receded from me. The intimacy of the gesture caught me off-guard, and my stomach dropped a little. Then the moment passed. I was starting to relax when there was a knock at the door. Before anyone could answer, a tall woman with dark wavy hair bounded into the room. "Noelle, this is our friend Trouble," Josh told me. "She's coming out partying with us tonight."

Soon we arrived at a spirited Georgetown bar called Mr. Smith's. A few of Josh's college buddies stopped by, and everyone took turns buying drinks and shots. Mr. Smith's had in its employ a gifted piano player with a marginal singing voice, not that it mattered when every patron was trying to drown him out.

"Sweeeeeeeet Caroliiiiiine! Buh-buh-buh! Good times never seemed so good. So good! So good! So good!" we thundered, pumping our fists in the air.

"I got this round," I said. "Who wants whiskey?" Josh and Trouble raised their hands, while Mo-Mo stuck to her vodka cranberry. The three of us tossed back the shots. Josh slammed his shot glass on the counter, planted his face between Trouble's sizable breasts, and shook his head vigorously while vibrating his lips, a gesture known as motorboating. Trouble unleashed a howl. Appalled, I glanced over at Mo-Mo but she was laughing with delight.

Okay. Now I remembered why we didn't work out.

We left after Trouble got kicked out for sticking her hand in the ceiling fan and nearly breaking it. We relocated to a nearby dance club, where Josh immediately began flirting with a group of middle-aged African American women. Soon they were all out on the dance floor where he was grooving behind one of them while repeatedly pretending to spank her ass. Mo-Mo watched the proceedings with amusement, occasionally snapping pictures with her cell phone. I excused myself to the bathroom. When I got back, Josh, Mo-Mo, and the rest of our group had been lost to the dance floor. I climbed up on a raised platform crowded with a group of giggly dancing coeds and spotted Monique encircled by some frat boy types. One of them was dry humping her aggressively. I scanned the room for Josh. He was break-dancing while the women from earlier cheered him on. I felt oddly protective of Monique. A rescue operation was clearly in order. I hopped down from the platform and danced my way over.

"Hey girl!" I shrieked, throwing my arms around Mo-Mo, sliding between her and the guy. She flashed me a look of relief.

"Yeah!" the guy cheered, thinking he now had two girls to dance with. He grabbed my hips and ground his pelvis into my butt. Perfect. I began by thrusting my pelvis wildly, so wildly, in fact, that I knocked him back several feet.

"Whoa!" he called out, unsure what to make of this dance floor interloper. "Some moves you got there."

He attempted to get behind me again and again, but I thrust back my pelvis hard so he just bounced off. When he tried to slip past me to Monique, I proceeded to phase two: a call to arms. I waved my arms enthusiastically to the beat, lashing out unilaterally against everyone within several feet. The guys backed away in alarm. After a few minutes, they realized they were not going to get anywhere and slunk off into the crowd.

Mo-Mo smiled at me. "That was great!"

When we got home at three A.M., it was Mo-Mo who set up the couch and made sure I had a glass of water and extra blankets. I took a sip of the water and gulped down my sleeping pill for the night. The second their bedroom door closed, I popped in the Jessica-recommended, sounds-of-sex—blocking foam earplugs. For at least an hour, I stared at the ceiling listening to my heartbeat, which somehow sounded magnified through the earplugs, a heartbeat in surround sound. I had three

major challenges coming up; despite all my worry time and compartmentalizing, it was getting harder not to stress about them. I was too
wound up to sleep. Also, I was on a couch, so it was harder to get comfortable. I squinted at the clock on their DVD player: 5:00 A.M. In a few
hours they'd be getting up and I'd be crabby from not getting enough
sleep. I didn't want to be a rude houseguest, right? I rolled over and
reached into my backpack, which was on the floor leaning against the
couch. I pulled out the bottle, reasoning that this didn't count because
I was doing it for Josh and Mo-Mo, not myself. They shouldn't have to
suffer because of my insomnia. Anyway, I didn't need much. Just half
a milligram. It was almost nothing.

The next morning I woke up to Mo-Mo handing me a mug of fresh
coffee and asking if I'd care for pancakes, which she'd made herself. I
was touched.

"Josh had to run to the office," she said as we settled around the
kitchen table, "but he'll be back soon to drive you to the bus station."

She and I chatted easily over breakfast. It was just the two of us,
Josh's past and Josh's future. I was amazed at how natural it felt and
how much I liked her. Out of all the challenges, this was the most personal. Most of my fears were about letting go of something that had
been holding me back. With Josh, I faced my fear of letting go of a person—or in this case, an *idea* of a person.

When Josh drove me to my bus, Mo-Mo came along. Riding in the
backseat of their station wagon to the bus stop, I felt as if they were my
parents. They hugged me good-bye, and I hoisted my backpack over
one shoulder and got in line for the bus. Just before I stepped on, I
turned and saw they were still standing in front of their car waving
at me.

I got a window seat and spent most of the ride staring out of it, seeing nothing. I'd been sabotaging my relationship with perfectionism.

I'd been searching for Matt's flaws, looking for clues that our relationship might not work out. In the process I'd taken his positive qualities for granted. I could see that now.

When I was an hour away, Matt sent me a text message. "Baby, you've almost returned to me!" it said. "Can't wait to see you!"

I'd needed to be reminded of the horrible arguments I had with Ben to appreciate that Matt and I didn't fight. Remembering how emotionally distant Isaiah was made me grateful that Matt wasn't withholding. He gave himself fully and freely to me. He spent hours driving me to activities that he had no vested interest in. At my trapeze recital he'd videotaped my performance and stepped on several audience members as he scrambled to get the best angle. I couldn't believe I ever doubted him. Josh and I had had our moments, but we'd also burnt out quickly. The passion had been unstable. Josh had often left me *wanting*. Matt's love may not have been showy but it was there, always. It was a well I could draw from at any time. When I was around Josh, I became my best version of myself. Josh dragged me out of my comfort zone and forced me to try new things. But I could become that person for myself. I *had to* become that person for myself.

"There's my girl!" Matt called out when I opened the door to his apartment. I followed his voice to the bedroom where he was reading one of those goofy science-fiction novels he's addicted to. "You were gone so long I forgot how sexy you are! I love this shirt on you," he said, pulling me down on the bed to nuzzle me. I was wearing a dirty Yale T-shirt. I looked at his face to see if he was teasing, but he was completely serious.

Three weeks after my visit I received an e-mail from Josh:

I wanted to let you know that Mo-Mo and I got engaged this past weekend. She was very surprised, but we are superexcited and looking forward to planning everything. We both wanted to share the news with you.

At the word *engaged* my heart instinctively leaped up in my chest. Then I read it over, acknowledged it, accepted it. Everything was as it should be.

"Wow. Congrats!" I wrote back. "How did you propose? Was it a big grand gesture or did you keep it simple?"

"I put together a scavenger hunt at the US Arboretum," he e-mailed, "and at the final clue, where I had first told her I loved her, we were getting ready for a picnic and I got down on one knee. So kind of a mix of the grand and the simple."

For a second I was speechless. Then I forwarded the e-mail exchange to Jessica and attached a note saying. "He stole my bit! For his engagement! Oh, that is cheap. That is *cheap!*"

"Well, think of it this way," she replied, "After the rabbi pronounces them man and wife, Josh will probably motorboat the maid of honor."

Chapter Thirteen

⁓

Courage is more exhilarating than fear and in the long run it is easier. We do not have to become heroes overnight. Just a step at a time, meeting each thing that comes up, seeing it is not as dreadful as it appeared, discovering we have the strength to stare it down.

—ELEANOR ROOSEVELT

"So next you're doing stand-up comedy, then you're going to work at a funeral home for a week, right?" Dr. Bob asked.

"Then Kilimanjaro. Remind me again why I saved the biggest challenges for the end?"

"It's too bad you couldn't have combined the two and done stand-up at the funeral home. You'd have been guaranteed a good crowd!"

I groaned.

"You know, public speaking is the number one fear in America," he said matter-of-factly. "Death is actually number two."

"Yeah, Jerry Seinfeld has a bit about that. Most people would rather be in the coffin than delivering the eulogy. I get that. When I get up onstage, death feels like a reasonable alternative."

"So stand-up is your biggest fear, then?"

I nodded. My muscles tensed up just talking about it. "I would rather do anything, anything else. I'd honestly give everything in my bank account—well, what's left of it—to get out of doing stand-up. It's completely irrational, I know."

"Not at all. People are afraid of public speaking for the same reason animals get nervous when they're surrounded by potential predators," he said. "Remember it all goes back to evolution. Our ancestors had dangerous neighbors. Public speaking involves taking a dominant position in front of strangers. Imagine doing that ten thousand years ago in a fierce environment surrounded by starving, angry, and somewhat paranoid strangers. The person who got up in front of the strangers and gave a speech ended up as dinner."

"So the audience is going to eat me alive, is what you're telling me?"

He winked. "If you're lucky, they'll kill you first."

Several weeks before, I'd e-mailed Chris to ask if he knew any comedy clubs holding open mic nights. As a blogger for *New York* magazine's website, he was plugged in to everything that was happening in New York. He'd e-mailed me a press release for something called the New York's Funniest Reporter contest, in which journalists were to take the stage at a comedy club and perform six minutes of stand-up. At the end a panel of judges would declare a winner.

"And it's for charity," he'd added in a follow-up e-mail. "A hundred percent of proceeds go to Operation Uplink, which buys phone cards for soldiers in Iraq so they can call their families." This convinced me I'd found the right venue to make my stand-up debut. Charity was such a big part of Eleanor's life, I'd been regretting that I hadn't done more to honor that this year.

During her years in the White House, from 1933 to 1945, Eleanor delivered an estimated fourteen hundred speeches across the United States and abroad, almost always from minimal notes. She also taught classes in drama, literature, and American history. Her public speaking career lasted forty years, and she's still known as one of the most popular orators of all time. She accomplished all this even though she started out petrified of public speaking, suffering from stage fright well into adulthood. She was pushed into it by Louis Howe, a former newspaper reporter and Franklin's chief political adviser. He was a chain-smoking gnome of a man who once declared he had one of the four ugliest faces in New York. While Franklin was recuperating from polio, Howe encouraged Eleanor to give speeches at political events to keep her husband's name in the air.

"I remember my own feeling that this was a thing I could not possibly do," Eleanor wrote in *You Learn by Living*.

"You can do anything you have to do," Louis insisted. "Get out and try."

She delivered her first address at the age of thirty-eight at a luncheon for the Women's Division of the New York State Democratic Committee. Louis sat in the back of the room and observed.

"I was a most unwilling victim. When I got up to speak I was shaking with fear because I had no idea how to prepare a speech, how to talk, how to handle an audience," she remembered. "When it was over he criticized everything I had done, particularly the fact that I had giggled every now and then, though there was nothing funny."

"Never write it down," Louis advised her. "You will lose your audience."

That advice was something of a problem for me. Because where I truly excelled was blanking out. I'd start a sentence and a few words in I'd forget where I was going with it. One time I was approached by a VH1 producer who was looking for pop culture writers to appear on a short segment they were doing about hot Hollywood couples. For the audition, they brought me down to the studio and led me into a small windowless room that contained a video camera sitting astride a metal tripod. It was the kind of space where you'd expect to have your head cut off on the Internet. There I sat down on a metal folding chair next to a producer, who asked me very basic questions like "What do you think of Jay-Z and Beyoncé?"

"Uh, Beyoncé is—" I stopped, my mind blank.

"It's okay. Just start again," the producer urged.

I began again, in a shaky voice, "The thing about Beyoncé and Jay-Z—" I broke off. "I'm sorry, I'm just nervous." Even being in a tiny room with just a producer and a cameraman was overwhelming. This went on for about ten minutes before the three of us unanimously decided that I should leave.

In the weeks that followed, I stumbled upon material from all sorts of sources. Jessica phoned me one night to announce a mutual friend of ours was pregnant.

"Can you believe it?" Jessica sighed. "Another one bites the dust."

"Does it make me a bad person that I'm kind of repulsed by the idea of pregnancy?" I asked. "A tadpole swims up inside you, latches on like a parasite, and grows bigger and bigger until one day it bursts forth? I'm sorry, that doesn't sound like a miracle to me. That sounds like something you picked up in Mexico."

"That's why I'm going C-section all the way. I don't negotiate with terrorists."

I laughed. "Did you seriously just compare babies to terrorists?"

"Anyone who takes your body hostage, explodes through it, and leaves behind that kind of path of destruction is a terrorist. And what's the protocol in a hostage situation? You send in a team to take the terrorist out."

Cradling my cell phone between my shoulder and cheek, I fished around in a kitchen drawer for a pen and notepad. "Wait, can you repeat everything you just said?"

On the subway one day a bodybuilder type with an aggressive suntan took a seat next to me. He immediately spread his legs into a wide straddle, committing several violations of personal space as his knees and meaty thighs invaded my leg area. This happened a lot on public transportation, and usually I responded by primly pinning my legs together and angling them away from the offender to minimize human contact. But this time, when I scooted over a few inches, Roids spread out even more as if to say, "Great! More room for me."

I turned to face him. "Excuse me, but this whole *situation* you've got going on here"—I gestured to his legs—"it's not working for me."

Roids looked at me in surprise. "What situation?"

"You with your legs spread out like you're home on the couch watching the game. And I'm over here riding sidesaddle in my own seat. I mean, what's that all about?"

He shrugged his beefy shoulders helplessly. "It's because of my balls, lady! I gotta make room for my balls!"

I arched an eyebrow. "Well, I gotta have room for mine, too, honey."

Roids laughed. "I like you. You got sass, lady." He brought his knees in until they were facing straight ahead and I relaxed my legs again. *What is it with men and their testicles?* I thought. *Guys act as if balls are celebrities and their thighs are bodyguards clearing everyone to the side. ("Out of the way, people! Give them some air!") Testicles are like the Olsen twins of anatomy—but with cleaner hair.* Suddenly, I had another bit.

"So when do I get to hear your jokes?" Matt asked a few days later. We'd just had dinner at our favorite French bistro and were walking back to his apartment, holding hands and swinging our arms.

"At the show." I'd considered testing my material on Matt, but I kept thinking of an essay I'd written the year before. I'd planned to send it to a newspaper column that published reader submissions but had asked him to edit it first. When I'd handed it to him, it was eighteen hundred words; when he'd handed it back to me, all but six hundred words had been crossed out. He'd cut so much that I lost the narrative thread and had no idea how to fix it. I'd also lost my nerve to submit the earlier version.

He dropped my hand and stopped in the middle of the sidewalk.

"What? I don't get to hear them first?" He looked so genuinely hurt that I relented.

I launched into my routine: "So porn star Jenna Jameson had twins this year. Which reminds me of this article I read recently in *In Touch* magazine. They were interviewing celebrities about their tattoos and Jenna confessed she once almost got a Hello Kitty tattooed on her wrist. But she decided not to because she thought, 'How would I explain that to my grandchildren?'" I paused. "*Really, Jenna? That's* what you're worried about explaining to your grandchildren?" I looked at him expectantly.

He didn't laugh and actually cringed a little.

"Oh come on! That's my best joke!"

After a few seconds, he said, "Maybe if you said it slower?"

"Okay, never mind! No more jokes for you."

Before bed that night I swallowed one sleeping pill and decided to throw in a half of another pill for good measure. As the contest approached, my sleeping pill intake had slowly started to rise. When I wasn't getting enough sleep, my head was foggy and I couldn't write my routine—or anything else. And I was also training to climb a mountain and couldn't exercise if my body was fatigued. It was a catch-22. I couldn't stop taking the pills because I had to train for Kilimanjaro, but I needed to stop the pills before I got to Kilimanjaro. I placed a pill on my kitchen counter and positioned a knife across the middle. When I pressed down, one of the halves went rocketing onto the floor. *Damn. They're going rogue now?* I should've taken this as a sign. Even the pills didn't want me taking them. Instead I got out a flashlight. Then I was on my knees, one side of my head pressed against the floor, sweeping the beam back and forth. The addict in me couldn't let that precious rogue half pill go to waste. Finally I saw it, nestled amid the dirt and a ball of lint beneath my fridge. I blew on it, rinsed it off, and popped it into my mouth. When I tipped my head back and swallowed, I saw my parakeets staring down at me.

"I know how this looks," I said.

Meanwhile, Matt's reaction spooked me enough that I put off practicing the routine until the contest was a week away. I couldn't even bring myself to say my lines out loud when I was alone in my apartment. This shouldn't have come as a surprise, I suppose. Procrastination is the lazy cousin of fear. "When we feel anxiety around an activity, we postpone it—whether it's doing our taxes, working on a project we're not sure we can handle, or having a painful conversation," Dr. Bob once told me. "You'll never feel ready. You have to do things now—even if you don't feel ready."

Now that I knew the two were related, whenever I caught myself putting something off, I looked for ways to make the dreaded activity less intimidating. If I was suffering from writer's block, I'd transcribe passages from my favorite books until inspiration struck. Sometimes it helps if you start with someone else's words. So I turned to one of my favorite comedians, Jim Gaffigan. I popped in my headphones and cued up his comedy album on my iPod. I grabbed my hairbrush off my bedside table and stood in front of the mirror. Performing Gaffigan's routine was less threatening because I wasn't judging the material.

"Am I the only one who finds it odd that heaven has gates? GATES?" I repeated along with Jim, mimicking his incredulous tone. "What kind of a neighborhood is heaven in? What, you die and go to a gated community? Are the gates really necessary? Are they like, 'Yeah, we got a lot of kids sneaking in and using the pool. Getting those gates wasn't easy. We had to go down to hell and get a contractor and everything.' "

When I was sufficiently loosened up, I tried a few lines from my first bit. My voice sounded high pitched and uncertain. Eleanor had this problem when she started public speaking. Her voice, which was high to begin with, got higher as she became increasingly uneasy. Instead of making her points forcefully, she'd trail off with a nervous giggle. A vocal coach taught her to lower her pitch during speeches.

High vocal tones suggest anxiety and shrillness while warmer, lower tones convey control and authority.

I cleared my throat and repeated the bit again, using a deeper, casual tone. Better. I got out my digital tape recorder I used to interview celebrities and performed my entire six-minute act. Then I played it back and listened. I rushed through my lines, as if I was trying to get as far away from them as possible.

I called Mark Anthony Ramirez, the comedy mentor assigned to me by the people who ran the contest. Each journalist was paired up with a professional comic who would field any questions we had about stand-up. Mark Anthony had performed at every major comedy club in New York so I trusted his judgment.

"The most important component of stand-up is projecting confidence when you're onstage," Mark Anthony said.

"But I'm *not* confident onstage!"

"Then pretend you are. Fake it till you make it. As soon as the audience senses you're not confident, you've lost them."

"Or maybe they'll laugh more out of pity?" I asked hopefully.

"If the performer is nervous, the audience is going to be nervous for that person. People laugh when they feel at ease, and people feel at ease with someone who projects an aura of self-control."

"So crapping my pants onstage is out of the question?"

He chuckled. "You'll do great. Oh, but make sure you know your routine backward and forward because when you get onstage, you'll be so juiced up that you'll probably blank out a bit."

Blank out? My public speaking specialty. Stand-up comedy was one of the hardest forms of public speaking because you couldn't just go onstage with a general sense of your talking points and wing it. In stand-up you had to get to the jokes in as few sentences as possible before the audience lost patience. Your routine had to be razor sharp. Forgetting one line or having even one word out of place could ruin a joke. Six minutes was a lot of material to memorize. Mark Anthony had me read my routine over the phone.

"I have to tell you," he said when I finished, "of all the people I've seen perform in this contest over the last few years, you're by far the most raunchy and in your face. It's risky. I hope you can pull it off."

I knocked loudly, struggling to be heard over the TV blaring so loudly that I could hear it in the hallway. Matt opened his apartment door wearing boxers with tiny bulldogs printed on them and the wire-rimmed glasses he only wore while lounging around on weekends.

"Hey! I thought you were at home rehearsing for the show!" His expression turned to concern. "Are you okay, honey? You look a little . . . overwrought."

I forced my mouth into a smile. "I'm fine!" I said, a little too brightly. "I just want to run an idea by you."

Matt was the most rational person I knew. If my conscience were to come to life in human form and be given a ridiculously good head of hair, it would look like Matt. If I could get Matt on board with my plan, then I could justify it to myself.

"Everyone is so excited to watch you perform tomorrow night," Matt said as I plopped down on his couch. "What time should we get there for good seats?"

How about never? I thought. *Does never work for you?* Ignoring the question, I asked. "Did I ever tell you when I decided to be a writer?"

He shook his head.

"I was a freshman in high school English, and our lesson that day was on satire. My teacher divided us into pairs and told us to work together to draft our own satirical essay. My partner was the class clown, Jon. He was busy goofing off, so I ended up writing the whole thing myself. When it was time to read aloud, Jon read one half and I read the other. The students were falling out of their chairs laughing at Jon's part, but my part fell completely flat. Shakespeare was right: all the world's a stage. But some of us are meant to be performers and others are meant to be writers. That's when I knew I was meant to be a writer."

Matt eyed me skeptically. "Where are you going with this?"

"It's just—" I avoided his gaze. "There's a reason they never cast me in the school plays and stuffed me in the back of the chorus—and it's not because I'm tall. There's a reason I'm bad on TV. I already know I'm not a good performer. Why do I have to get up there and show everyone else, too? What will that accomplish? Maybe the point of this project is that I don't conquer all of my fears, but I accept my shortcomings?"

His expression softened. "Baby—"

"There's even an Eleanor quote about this." I removed a piece of paper from my back pocket and read, " 'Perhaps one of the most difficult things any of us has to do is to be able to say clearly, "There is a limitation in me. Here is a case where, because of some lack of experience of some personal incapacity, I cannot meet a situation." ' So, you see, I think Eleanor would take my side here—Jesus, could that TV *be* any louder?" I said, turning toward the television set across from us. I snatched up the remote and stabbed at it with my fingers, looking for the power button. Instead, the screen turned to snow and the crackling sound of static roared from the TV. "Oh, just make it stop already!"

Matt gently pulled the remote from my hand. With one press of a button the TV darkened and silence filled the room. Tears sprang to my eyes, and he became a kaleidoscope of Matts swirling before me. "I don't want to do this!"

"I know it's scary." He reached for me. I collapsed against him and let him hold me while I bawled. It was one of those shuddering cries that bordered on hyperventilation.

"Haven't I put myself out there enough this year?" I gasped between sobs. "Haven't I embarrassed myself enough and taken enough risks? Can't I keep this one last shred of dignity? I'm exhausted, Matt. I'm tired of being scared all the time."

"I know you are, but twisting Eleanor's quotes can't get you out of this. Only you can. But personally, I don't think you should. Think of all the scary things you've done this year. Were any of them as bad as you thought they'd be?"

I sniffled and wiped my face. "No."

"In fact, you enjoyed them, right? Weren't you just telling me that, with the exception of shark diving, you'd do every one of them again?"

"Yes," I admitted grudgingly, resigned to my fate.

He smiled down at me and stroked my back. "I believe in you, honey. You went skydiving, remember? This can't be scarier than skydiving."

"Yes, it can," I mumbled against his chest. "In skydiving you can only die once."

Comic Strip Live was crammed in the middle of a miscellaneous block of delis and bars on the Upper East Side. The walls of the foyer were covered in framed black-and-white headshots of comedians who'd performed there. Among the ones I saw when I walked in: Jerry Seinfeld, Eddie Murphy, Chris Rock, and George Carlin.

The main room smelled vaguely of cigarettes, though smoking in New York bars was outlawed years ago. At the front, a circle of light formed a bull's-eye over the stage. I was surprised by its smallness. It was a glorified step, maybe five feet wide. Except for the strip of faux brick wall behind the stage, the room was painted crimson. *So you can't see the blood on the walls,* I thought.

Earlier that day I'd asked Chris—who did improv comedy in college—if he had any last-minute advice. "If not that many people show up, when you get onstage ask the audience, 'Hey, who brought in all these chairs?'" I definitely wouldn't need Chris's line. There were about two hundred people jammed around the various rectangular tables.

There were to be ten performers that night, all of us in either print or TV journalism. We were corralled in a side alcove, separate from the audience. I introduced myself to the others and took a seat in a red leather banquette in a quiet corner where I could go over my routine in my head. I was having a love-hate relationship with the passage of time. Every minute that brought me closer to my biggest fear also brought me closer to the *end* of my biggest fear.

In the club's dim lighting I could see Matt had staked out a corner with about thirty of my friends—old work colleagues, college buddies, Matt's college buddies, and of course Chris, Jessica, and Bill. The sight of them gathered here in support was so moving that I had to blink back tears. At the same time, my heart quivered, knowing that if I bombed, it wouldn't be in front of a group of strangers I'd never see again.

A heavyset Latino guy wearing a black leather jacket, a baseball cap, and small gold hoop earrings came over and held out his hand.

"Noelle? Mark Anthony. Nice to finally meet you in person. You're third in the lineup, by the way."

"Good, I'll get it over with quickly," I said, while simultaneously texting Matt the number so he'd know when to expect me. "I don't know why I'm so nervous. It's only six minutes of my life, right?"

"Five, actually," Mark Anthony said.

I looked up abruptly from my BlackBerry. "What?"

"Yeah, sorry about that. Apparently I was wrong when I told you your routine should be six minutes. The emcee just told me you guys only get five minutes."

I was momentarily relieved. One less minute I'd have to be onstage. Then I realized what this meant.

As if reading my thoughts, Mark Anthony said, "So you'll need to cut two bits from your routine. Want anything from the bar?"

I shook my head, dumbly, unable to speak. My mind raced. The whole routine had been carefully constructed with seamless transitions from one bit to the next. It had taken me a week to memorize it. Cutting two bits would throw everything else off. I reached into my purse and pulled out my routine, neatly typed up on three sheets of computer paper. I scanned it quickly, frantically flipping the pages. I could play it safe and cut the two most vulgar bits. At least I wouldn't have to worry as much about offending people.

"Does anyone have a pen?" I called out desperately.

A passing waitress extracted a blue ballpoint from her apron and handed it to me. Noting my panicked expression, she said, "Keep it, honey."

I went to strike a line through the vulgar bits, but when I pressed the tip of the pen into the paper, I couldn't bring myself to do it. Even though they were the most risky, they also had the potential for the biggest laughs. Before I lost my nerve, I turned the page and struck out my bit about baby terrorists and the one about guys sitting spread-eagle on the subway.

The emcee, a cute dark-haired guy in his thirties with a faint Brooklyn accent, took the stage to warm up the crowd. All the while I was feverishly scrawling new transitions to fill the holes left by the deleted bits. A few minutes later he introduced the first comic, a political reporter for the *New York Post*. His opening line: "Okay, I'm a Catholic, West Indian black Republican. Anyone else here?" Silence. "Ah, I thought not." The crowd guffawed.

I tried to practice my routine, but I only got as far as the opening line and then stopped. Oh my God, I wasn't even onstage and I was already blanking out! I was clutching my routine so hard between my hands that the pages were puckering. These sheets of paper were my safety net. If I took them up there and read from them, I'd be sure not to have any awkward blunders. I'd also basically disqualify myself. Everyone else had their routines memorized. There was no way I could win if I used my notes. A producer for CNN was up now. Her delivery was incredibly deadpan and she opened with, "Can I ask you guys a question? Is it wrong to gorge yourself on pizza and ice cream while watching *The Biggest Loser*?" Her routine was getting a lot of laughter from the audience, but I could barely hear over my breathing, which was steadily increasing in both frequency and pitch.

Mark Anthony returned from the bar swigging a beer and gave me a friendly punch in the arm. "You're up next, kid."

Dr. Bob had once told me that when animals and humans feel threatened, we experience an adrenaline rush. Adrenaline is a performance enhancer, there to give us the energy to either dive in and persevere or escape. How we react is known as the fight-or-flight response. I chose flight.

"I can't do this," I whispered to Mark Anthony. "I'm not ready."

He looked at me for a moment while rubbing his goatee. "Okay, I think I can move you down in the lineup and buy you some time."

"Move me. Please." He hurried off to alert the emcee.

Something that also happens during the fight-or-flight moment is your body reroutes the blood in your stomach to your muscles, where it is most needed. The feeling of blood being pulled from your stomach creates the fluttering we call butterflies. Digestion temporarily shuts down, which can cause diarrhea. Let's just say, I very much needed to visit the ladies' room.

When the emcee announced a *Good Morning America* reporter as the third contestant, Matt shot me a confused look from across the room.

"What's wrong?" he mouthed.

I shook my head as if to say *I don't want to talk about it* and looked away. Mark Anthony returned a few minutes later.

"Want to go next?"

I put up my palm. "Still not ready." Ten minutes later he asked again.

Without looking up from my paper, I answered, "Nope."

"Noelle, look at me," he said sternly, but when I looked up, his face was kind. "I can push you down to seventh, but that's it. You just have to get up there and go for it."

Next up was an indie magazine writer who strode onstage in a pink rock star wig. She was wearing an elf costume and a strapless sequin dress simultaneously. She leaped around the stage and growled at the audience, unleashing an onslaught of groaners like:

"Knock knock?"

"Who's there?"

"Angry cow."

"Angry cow, who?"

"*MOOOOOOVE!*"

And that was one of the good ones. The audience sat in stony silence, punctuated by the occasional heavy sigh. I breathed deeply to slow down my heart rate. I remembered something Dr. Bob once told me about a poem called "The Guest House" by a thirteenth-century Persian poet named Rumi. In the poem, Rumi suggests we think of our emotions as guests who show up at our house unexpectedly. No matter who shows up—joy, sorrow, depression—all should be welcomed inside and entertained because there's a lot we can learn from them.

"Think of your fear as a guest who shows up unannounced. Invite the fear into your guest house and listen to it. No matter how much he whines, eventually you'll find you can just tune him out and keep going about the business of your day. Accept him, be kind to him. If you make friends with fear, you won't mind when he shows up at your door. You may even look forward to his visits."

And so I invited fear into my house. I listened patiently while he bitched and moaned. "What if you forget what to say? You'll look like a total idiot. All of your friends will be embarrassed for you. They'll pity you. What if the older people hate your dirty jokes? What if you talk too fast or hold the mic wrong so no one can hear you? Remember, you've never held a mic before!"

"Anything is possible," I replied calmly. "But I think it will turn out okay. It usually does. And if it's really bad, at least it'll make for a good story one day."

Soon I noticed that fear was repeating himself. At first I'd thought there was a lot to fear. But he was just rehashing the same few worries over and over. I stopped feeling overwhelmed and I relaxed. I even tuned him out, as Dr. Bob predicted. Fear is boring. After fifteen minutes, he quieted down so much that I didn't even hear him slip out the door. I knew fear would be back again. He'd continue to drop by unannounced, probably for the rest of my life. But after spending so much time with him this year, I understood him better.

Suddenly, it was upon me. The emcee announced: "She's a freelance writer who's been published in magazines like *Rolling Stone, Maxim,* and *Us Weekly.* Please give a warm welcome to Noelle Hancock . . ."

Mark Anthony clapped me on the back. "Get up there and make that stage your bitch!"

I folded up the papers and shoved them in my back pocket. As I awkwardly wove my way to the front of the room, turning sideways to squeeze between the tables, Matt caught my eye and gave me a thumbs-up. Chris and Jessica waved eagerly, big grins on their faces.

Bill shouted, "You got this, Hancock!"

When I stepped onto the stage, my friends vanished into the darkness along with most of the audience, obscured by the glare of the spotlight. Only people sitting a few feet away in the first two rows were visible; mostly contestants' parents in their sixties. There were many tired faces, one hand pressed to a cheek, propped up by an elbow on the table. A few frat guys sat tilted back in their black wooden chairs, arms folded across their chests.

I drew the mic under my chin and tried to look confident. "Is everyone enjoying the weather?"

The silence in the room was absolute. The audience was tired of participating. The pink wig girl had worn them out with all her questions. Finally someone—possibly Matt—hooted in confirmation.

"Me too," I continued breezily. "I love the fall because I hate bees—and fall is like their off-season. Because in the spring and summer, you never know when someone's going to turn to you and say: 'Don't move! There's a bee on you. Seriously, don't move!'" I made a confused face. "Don't *move*? It's, like, whose side are you on? You want me to hold still so the bee can find a good vein?" The crowd laughed. Emboldened, I plowed on.

"You ever wonder who discovered honey? I do. Because I wanna meet the guy that looked over at a beehive and thought to himself, 'You know what would be a really great idea? You know those dangerous insects over there? I'm gonna break into their house and steal all their shit!' That's like breaking into the headquarters of the Latin Kings." I'd been worried not everybody would've heard of the street gang, but this received a good amount of laughter from the audience. "Bees are really just tiny gang members, right? Always rolling with their crew, always packing a weapon. Bees don't even have butts—they have *daggers* where their asses should be. The weapon is built in! So don't tell me 'don't move.' Because that's not a bee sting, that's a fucking drive-by!"

I could hear Chris's punchy laugh amid the cheers and applause. I paused and silently counted to three, which Mark Anthony had said was another way to signal the crowd that I was about to change topics.

I drew a deep breath, wondering if I was about to offend everyone in the room. "So I don't like giving blow jobs—"

The laughter hit like a gale-force wind, stopping me before I could even get to the punch line. It was so startling I almost took a step backward. The tone of the laughter was a mixture of surprise from the men and a kind of admiring identification from the women. When the room quieted down, I started again.

"So I don't like giving blow jobs for the same reason that I don't like people projectile vomiting into my mouth."

The response was even louder this time. The room seemed to undulate. People were actually doubling over, I realized with amazement. I put my hand on my hip and gave the audience a cocky half smile. They whooped in approval. I glanced down at the first row and my smile faltered. The parents were scowling up at me. One of them grimaced in disgust. Clearly, I'd lost them after the honeybees. *Shake it off,* I admonished.

"Blow jobs are the reason girls start having sex in the first place, right? It's more of a gateway sexual activity for us." I leaned over the microphone as if it were a penis and held my hair back with one hand.

"What happens is somewhere around the third one we stop and we think to ourselves, 'Well, this is some bullshit!' " Waving my microphone/penis, I continued. "And I know someplace else I can stick this so that I can actually *breathe*—and watch TV at the same time."

People were howling. Howling! I paced the stage, playing to the crowd.

"My high school friends and I lost our virginity around the same time—because we were followers as well as sluts." Loud chuckling. "And of course we talked about it afterwards because girls have to talk about everything. All my friends said, 'Oh my God, it hurt so much! Didn't you think it was painful?' And I remember thinking, 'Painful? Really? Um, that wasn't *my* big takeaway from the situation.' " Then, like an afterthought, I added, "Then again, my uncle has a pretty tiny penis. Maybe that had something to do with it."

The incest joke drew a few horrified gasps from the front row, but they were quickly drowned out by laughter from the rest of the room. For some to love you, others have to dislike you—that's the nature of performing.

Everyone except the front row loved the Jenna Jameson joke, as I knew they would. While the audience was cracking up, I reached in my pocket and pulled out my safety net—the typed-up routine. My closing bit was long and there was a high probability of flubbing a sentence or two. I stared at the square for a few moments, debating. My routine had been flawless so far. If I read this last part from my notes, I was guaranteed the rest of it would be perfect. My fingers began unfolding the pages.

No.

Suddenly I stopped and stuffed the square back in my pocket.

"I'm going to leave you all with my craziest subway story. And this is absolutely true because there's no way I could make up something this twisted. One night I'm in a subway station waiting for the train. When it arrives, the car that stops in front of me is empty. And that should've been my first clue something was wrong because that never happens in New York. I step into the car and it smells like feces." The New Yorkers in the room snickered knowingly. "And that's because there actually *is* feces smeared on all the seats in this section." The laughter grew louder.

I continued: "Then I realize that the car isn't actually empty. It's that all of the passengers are huddled at the opposite end, clinging to each other for dear life. So I join them—because nothing brings people together like the fear of human excrement. At the next stop this couple gets on the train dressed in black tie, drunk out of their minds—it's like six P.M., by the way—and they go to sit down." A few groans of recognition went up around the room from the people who saw where this was headed. "We realize what's about to happen, so all the passengers start screaming, 'NOOOOO! STOP!' But the couple is hammered, so they don't understand what's happening and . . . they sit in the poo." There was a collective moan from the room, but it was mixed with laughter.

"So we're yelling at the couple, 'Get up! Don't sit there!' and they're slurring, 'Whaaaaaa? Whassss happening?' Finally they drunkenly stagger across the aisle to change seats, but since the whole section has been defiled, they sit in the crap a second time." This garnered shrieks of horror from the ladies while the guys clapped their hands a couple of times in that way that's reserved for reveling in another person's embarrassment.

"Now we're fucking losing our minds, saying, 'NOOOOO! Don't sit there either! Move!' So the couple slides over, smearing themselves across, like, five more seats of feces. Seriously, I haven't seen people rolling in shit like that since I watched German porn." The frat boys in the front were practically convulsing at this line.

"At the next subway stop, the couple gets off and heads off to their party in their formal wear, totally unaware that their backs look like a Jackson Pollock—if Jackson had had a feces period—and as soon as the doors close, every passenger bursts out laughing. Like, *howling*, pounding-each-other-on-the-back kind of laughter. It was beautiful. Something brought a group of strangers together in New York that night." I paused for effect. "And it was the shit."

A brief silence descended and then the crowd erupted.

"Thank you so much!" I yelled over the applause and cheers, returning the microphone to the stand. "You've been a wonderful audience!"

Unbelievable. The rush that came over me was like nothing I'd ever experienced. It was a joy so pure and profound and untainted that it was almost a high. I was in love with this moment. I wanted to pack up my belongings and move into this moment. I wanted to live here for the rest of my life. I was magnificently and disgustingly happy. And more than that, I was marvelous to myself.

As I made my way back to the comedian alcove, there were many high fives and backslaps from the other comics. "Well, this competition is over," one of them said with a grin. Another one leaned in and said in my ear, "You've got this thing locked down. Congratulations!" By the time I reached my seat, my phone was already buzzing with text messages from my friends.

Chris: "OMG, we are ALL blown away!"

Jessica: "I forgive you for cutting the baby terrorism. Because seriously you were AMAZING! You looked so comfortable up there. Your delivery was incredible!"

Bill: "UR kinda my hero. P.S.: I've never felt more uncomfortable than when that offspring of Pink and Rumpelstiltskin came onstage. WTF? Just a bad moment for everyone."

But my favorite message was from Matt. He wrote: "That was the most amazing thing I have ever seen you do. Or anyone do. You could be a real stand-up comic. It was that good. If you don't win, I'm setting this place on fire."

I didn't stop beaming throughout the next three performances, but I wasn't even listening. My God, I was really going to win this thing. I couldn't believe it! This would be my proudest accomplishment since getting into Yale. In my head I rehearsed a quick acceptance speech. How amazing to get to thank Matt and Chris and Jessica and Bill publicly for their support! Maybe I'd even thank Eleanor. I'd briefly describe the project and how I'd put this fear off until the end because it was the scariest and then bring it home with some closing statement along the lines of "Dreams really do come true!"

When the last comic took his leave, the emcee appeared onstage. "And now for the winners of the 2009 New York's Funniest Reporter contest!" he said. My stomach was churning with excitement and nervous anticipation.

"Third place goes to . . ." He announced the name of the *Good Morning America* correspondent, who smiled beatifically and waved at the crowd.

"Runner-up goes to . . ."

It's extremely rare but there have been a few times in my life where I've received a precognition about something that's going to happen. It's not a suspicion but a certainty. And suddenly I knew that I hadn't won the contest. So sure was I that they were about to announce my name as the runner-up that I said it in my head along with them:

"Noelle Hancock!"

Even though I'd known it was coming, a tingling disappointment flooded through my body, as though my insides were blushing. My friends clapped hesitantly, unsure of whether they should be cheering for the runner-up position when they'd been so certain I would win. I smiled hugely, but my throat was throbbing from holding in my emotions. There was a stinging behind my eyes. *Do not cry,* I scolded myself.

The deadpan CNN producer was announced as the winner. She didn't look at all surprised by the win as she took the stage, and she didn't seem to really care. She accepted the award with a casual "Hey, thanks, guys." Then before departing she said, "Let's give a round of applause to the other reporters. You guys were *awesome!*" Then the show was over. Matt and my friends were across the room, but I walked toward them slowly, trying to gather myself before I reached them. I was mercifully intercepted by the emcee, who led me and the other two winners to the stage for ten minutes of pictures. Later, I'd look back at these photos and marvel at how two different features on the same face could express such conflicting emotions at once. My mouth was set in a wide grin, while eyes were dispirited and shiny from the unspilled tears still lurking behind them. When we were finished, I stepped off the stage and Jessica caught me in a frenzied hug.

"You were totally robbed!" she said. "I'm sorry I can't stay. I have an early flight tomorrow and I still haven't packed." In a surprising move, she was going backpacking for two weeks in Argentina.

"Thanks for coming! Now get out of here, you fucking hippie!" I tried to sound jaunty, but my voice came out in that desperately bright tone people adopt when they're trying not to betray their disappointment.

Then Matt was at my side, making no attempt to disguise his mood. His expression was pure indignation. "That was bullshit!" he declared. "I can't *believe* you didn't win." Then he brightened. "I can't get over how good you were up there!" He put his arms around me, linking them behind my lower back. "You were so confident, honey. It was like watching you come into your own. I've never seen anything like it."

When he said this, I realized how completely absurd I was being. My goal when I'd come here tonight had been to not forget my bits. Yet suddenly I'd raised my standards until nothing less than perfection would do. In doing that, I'd created my own disappointment out of something that should have been joyous. It was a habit I'd spent the last year trying to break. And right then, I made the decision to release my disappointment. I refused to be an enabler of my own unhappiness. There are enough real misfortunes in life to get upset about. With that, I exhaled and let the disappointment flow across my tongue and out of my body. I was surprised at how easily it faded away. What remained was not the giddy elation I'd experienced earlier but something more mature—a deep, gorgeous contentment that had settled in my very bones. I understood why this was different from my other fears. Sky-diving, flying a fighter plane, shark diving—these were things I hadn't *wanted* to do. Making a crowd of people laugh for five minutes straight was something I'd wanted to do but hadn't believed I *could* do.

That night I lay in my bed, grinning into the darkness, playing those five minutes in my head on an endless loop. I'd stayed up till five in the morning, not because of insomnia, but because I didn't want the night to end.

Chapter Fourteen

A great deal of fear is a result of just "not knowing."
We do not know what is involved in a new situation.
We do not know whether we can deal with it. The
sooner we learn what it entails, the sooner we can
dissolve our fear.

—ELEANOR ROOSEVELT

My whole life I'd worried about death. When I was a child, I went to bed every night expecting to be murdered. I'd watched enough episodes of *Unsolved Mysteries* to know that the world was full of people wanting to kill me, and what better time to kill someone than when she's asleep and you have the element of surprise? I developed a lengthy pre-bedtime ritual. I checked the locks on the windows, searched the closet to make sure no killers were hiding there. Then I looked in the laundry hamper because the closet's way too obvious. Most likely, anyone small enough to fit in a hamper couldn't be much of a threat, but when you have the element of surprise working in your favor, anything is possible.

Even as an adult, I imagined my death as an extraordinary turn of events. I'd be in an elevator when the earthquake hit and the cable snapped, sending me plummeting to the basement. I'd be lolling on my back in the ocean and feel a tug on my arm. When I'd turn to look, nothing would be there. Except the shark rearing out of the water to finish me off. There would be enough time to process the terror of what was happening, but no time to say my good-byes or ask someone to throw away my vibrator so my parents wouldn't find it among my personal effects.

My fear had nothing to do with the afterlife, but with death itself. All fears were rooted in death, Dr. Bob explained. "The strongest basic instinct is survival. Evolution has programmed our fears into us to help keep us alive." I couldn't give fear its due, I decided, if I didn't face my fear of death. The idea of working at a funeral home actually came from Dr. Bob.

"It seems to me you've already faced these death scenarios that you've imagined all your life, in almost every way you could," he said. "You've swum with sharks and jumped from planes. Now maybe it's time to try something different. Perhaps you need to go look some dead people in the eye, so to speak, and see what that feels like."

He had suggested that, instead of making this just a one-day event, like most of my challenges had been, I spend some real time with this fear—a few days at least. That way I couldn't just psych myself up until it was over, I'd actually have to try to get comfortable in the situation. I'd see every stage of the process, from the time the body was brought in, to seeing their families grieving at their funeral. It would be a form of what he called *exposure therapy*.

"Being around the dead will remind you that there is nothing to fear from the dead," Dr. Bob said, "and remind you to get on with the living."

The funeral home was run by a man named Terry. I'd found him through a friend of mine whose mother was an undertaker and had worked with Terry a number of years ago before he opened his own funeral home in a small town in Ohio. When I'd explained the project to Terry, he said I was welcome to come spend a week shadowing him and his employees.

The funeral home actually looked like a home, a two-story redbrick house with a porch on the upper level with rocking chairs, and bright flowers lining the path to the front door. A sign said PLEASE COME IN, and when I opened the door, a soft *ding* sounded inside. I stepped into what looked like a dining room with assorted brass lamps and red valances hanging regally around the windows. In the center of the room was a glossy rectangular table with matching wood chairs, the kind families gathered around on special occasions, except here there was always one person missing. An office door opened and a jolly-looking man in wire-rimmed glasses strode out.

"You must be Noelle!" And this must be Terry, the funeral director. He was tall but extremely pear shaped, as though the upper and lower halves of his body came from two different people. He'd told me on the phone that he was thirty-eight, but with his graying hair and little-boy face, he looked both older and younger than his years. The effect was incredibly endearing.

"Thank you again for letting me help out this week," I said, shaking his hand.

"I'm a firm believer in facing one's fears, as well as in educating people about the funeral directing business, so you'll have full access this week. Let me show you around."

I felt like I was about to walk through a haunted house. There could be a dead body lurking in any of these rooms. He ushered me through a small utility room just off the dining room and stopped in front of a closed door.

"This is our preparation room," he said, reaching for the knob. "It's where the bodies are embalmed and prepared for the funeral." He opened the door and I braced myself for the horror to come. But when the fluorescent light came on, it looked more like a regular lab than a mad scientist's lair. The walls were lined with cabinets holding various instruments and chemicals. All five of the stainless steel tables were empty.

"It was a slow weekend, so there are no bodies or services here today," he explained and led the way to another room just off the foyer. "Obviously, this is our casket showroom." A dozen gleaming caskets were parked end to end, like new cars at a dealership, lids propped open like hoods. Their satiny interiors and plump pillows were patiently awaiting their next customer. It was an eerie sight. Open caskets have always disturbed me. The way the casket was bisected and only the top opened reminded me of that magic trick where they pretend to saw a person in half. Open casket funerals, to me, possessed the silent horror of a magic trick that had gone horribly wrong and killed the person after all.

He took me up the stairs, which were covered in the same floral carpeting as the first floor, to ensure a light tread. The second floor hosted a chapel and a sitting room full of formal wingback chairs and stiff couches decorated in calming pastel blue and cream. "Amazing Grace" played softly on the funeral home's sound system. The entire CD was devoted to "Amazing Grace," I realized, each version played on a different musical instrument. "Amazing Grace" on piano, violin, harp, and so on, in an endless loop. This was the scariest part of my day so far.

I picked up a pamphlet titled *Coping with Death* from the handsome secretary.

"Terry, are you up here?" A voice called from below, followed by the drumming of feet running up stairs. "I'm gonna run over to the crematorium and—oh, sorry." A young man with slightly unkempt brown hair bound into the room but stopped short when he saw me. He wore black slacks and a white button-down shirt like Terry, but no sweater vest or tie.

"Noelle, this is my intern," Terry said. "Lucas, this young lady will be helping us out this week, trying to overcome her fear of death."

"Cool." He smiled goofily at me and pushed his sixties-style glasses up the bridge of his nose. "Hey, I'm taking a body to get cremated. Wanna come?"

I didn't, but this was what I'd come here for, after all. I looked at Terry and he waved us on. "Have fun, you two!"

Lucas led me out the back door to a six-car garage behind the funeral home, which was packed with limos and hearses in blacks and muted grays. Lucas made his way past them to the cadaver refrigerator, which was the height of a normal refrigerator but about seven feet deep. He grabbed the handle of the steel door, paused, and looked back at me uncertainly.

"You're not going to, like, faint, are you?"

"I don't know," I answered nervously. Looking for street cred, I added, "I've been to two open casket funerals."

He nodded thoughtfully, satisfied with that, and yanked open the door. I stiffened, having no idea what to expect. Lucas slid out a corpse on a gurney.

"Meet Mr. Danbury."

I exhaled, and my insides unclenched. The body was covered by a sheet except for a pair of feet sticking out at one end. They were a surprisingly normal color. More ripe peach than the cliché gray I'd expected. Lucas reached under the sheet and lifted up the man's fingers.

"See how they're turning purple?" he asked. "That's the beginning of the decomp process."

There was a tingling in my stomach as I eyed the dead man's mottled fingertips. I was relieved that, instead of taking the sheet off, Lucas rolled Mr. Danbury toward a van with the funeral home's logo printed on the side. Lucas was only five feet seven with a slight build, but he loaded the gurney into the back with ease.

Before climbing in the passenger's side, I held up a clip-on tie someone had left on the seat. "Is this yours?"

"Guilty," Lucas said cheerfully. He grabbed it and tossed it onto the dashboard. "I wear it when I'm picking up a body from a hospice or someone's house. Terry says it looks more respectable." Until I heard Lucas's drawl, I didn't realize people from rural parts of Ohio have southern accents.

He waited for a sedan to pass and slowly eased the van out of the driveway onto the two-lane country road. Cornfields glided by in a slow, hypnotizing blur. When the corn gave way to farmland, I snapped out of my daze to see two young boys in black vests, slacks, and straw hats running across a field. They were giggling, trying to keep up with a horse-drawn buggy clip-clopping down the adjacent dirt road.

"What's that about?"

Lucas glanced out the window to where I was pointing. "Amish. We got the highest concentration of 'em in the country. Decent folk. Pain in the ass getting caught behind one of their buggies, though."

"Have you always wanted to be a funeral director?"

"Always!" Lucas said proudly. "When I was a kid, I used to bury my sister's Barbies in boxes in the backyard. What do you do?"

"I'm a freelance writer. I write about pop culture, celebrities, that kind of thing."

"Have you ever interviewed celebrities?"

"Sure, all the time."

"And they pay you for that?" he asked in a high-pitched, disbelieving voice. "Wow, I can't wait to tell my friends what you do for a living. They'll never believe it."

Funny, I was thinking the same thing.

In his excitement, he didn't notice the sedan in front of us had stopped to make a left turn. He stomped on the brakes. Mr. Danbury and his gurney rammed into the back of the driver's seat with a loud *tha-rump*.

"I hate it when that happens," he said, massaging his whiplashed neck with one hand.

"Are you okay?" I asked, unsure of whether to be concerned for him or scandalized that he was just rear-ended by a dead body. Then something else caught my eye.

"Lucas!" I exclaimed in disbelief. "How can you work at a funeral home and not wear a seat belt?"

He shrugged. "I forget."

The crematorium was an unassuming warehouse tucked discreetly at the end of a long driveway. As we pulled up, my heart jolted against my chest, unaware of what I'd find inside. I prayed I wouldn't become sick.

"That's Fred, the crematory operator," Lucas said, nodding at a gray-haired man in overalls who appeared in the doorway. The man tipped his baseball cap in greeting.

Lucas hopped out. "Fred, this is Noelle. She's helping us out this week."

"Right pleasure, ma'am," Fred said with a country accent even heavier than Lucas's. When we shook, I thought about where his hand had been.

"What ya got for me today, boy?" Fred asked as Lucas opened up the back door of the van.

"Dude in his seventies. Died of some kind of cancer called"—Lucas referred to his paper, sounding out the word slowly—"my-el-oma."

They unloaded the gurney and wheeled it into the warehouse, stopping in front of one of the cremation ovens, which looked disturbingly like a pizzeria oven. It was made of special fire-resistant bricks, with a small steel door, just large enough to accommodate one body at a time. The metal tube on top was the exhaust chimney, which had a special ventilation system to remove the smoke and human odor from the cremation process. This was why the building smelled like aluminum and concrete, not burning flesh.

I thought of something Bill said to me once. I'd asked him how he wanted to be buried and he'd said, "I'd like to have my ashes compressed into diamonds and make all my friends wear me."

Lucas pulled off the sheet and I saw Mr. Danbury for the first time. It wasn't frightening at all, surprisingly. Instead, the sight of him pulled at my heartstrings. Dressed only in boxers, he had the defenseless look of a mannequin between costume changes. In fact, with his sleek peachy skin there was something almost nonhuman about him, like someone trying to do an impression of a human.

Fred sized him up. "Any jewelry or pacemakers?"

"Oh, thanks for reminding me!" Lucas pried the man's wedding ring off his finger and dropped it in his breast pocket. "I'll make sure that gets back to his children. He died in a nursing home only a few months after the wife. Kinda sweet, right?" He pressed his hands on the man's chest and felt around his breastplate.

"You can feel the pacemaker from the outside?" I asked.

"Yes, ma'am," Fred answered, "and if they have one, we have to cut it out 'fore they go in the oven or it'll explode right outta their chest. Hell, we still got one stuck in the side of the wall in there." He opened the oven and pointed inside at a small metal disk half embedded in one of the bricks, as if a miniature flying saucer had crashed into it.

Fred chuckled. "Just about scared the bejesus outta me when it went off. Sounded like a gunshot."

Lucas pronounced Mr. Danbury pacemaker-free, and he and Fred zipped him into a body bag and loaded him into the oven. The bag would eventually be vaporized, Fred explained, but was there for sanitary reasons. He checked the temperature gauge. The remains would be incinerated over the course of three hours at 1,700 degrees, then cooled down for one hour before they could be removed.

"You're just in time, matter of fact. This guy's about done."

I followed Fred around the corner to another oven. He opened the hatch and when the residual heat hit my face, I felt the millions of pores on my face expand simultaneously, like mouths opening to scream. Inside the oven sat a pile of ash and bones. The heap smelled of chemicals and gases I couldn't place. Using a long-handled wire bristle rake, Fred scraped the steaming pile into a tray and carried it over to a worktable. It was like watching a cooking show where the chef prepares a meal, sticks it in the oven, then immediately opens another oven and brings out the same dish, already cooked. The thought of this made me half gag, which I played off as a cough.

Fred explained that the bone fragments would eventually be fed into a machine that pulverized the bones into "cremains." "But before we pulverize 'em, we gotta dig through the bones and pull out all the manmade stuff." He picked a bucket off the floor. "Look here."

It was full of jumbled hardware—screws, metal pipes, nails—only they were scorched and covered in what looked like dust. With sudden queasiness, I realized I was looking at reconstructed joints, surgical pins, and metal limbs. And that dust was people. *People dust.* Using his bare hand, Fred plucked an ash-covered ball-and-socket mechanism off the top. My gag reflex wobbled again.

"That there is a hip replacement." He rummaged around like a kid going through his Halloween candy. "And this is my favorite!" Beaming, he pulled out a delicate lattice of thin metal tubes held together with screws. "This was in someone's spine, if you can believe that!"

I caught Lucas's eye. Noting my stricken expression, he cleared his throat and said, "We have to be getting back. Fred, I'll be back for Mr. Danbury's cremains tomorrow."

Fred dropped the lattice back in the bucket and blew the powder off his fingers. *Poof!* It formed a cloud—a person cloud—suspended in the air, then vanished.

"Nice meeting you!" I said, hurrying back to the van before he could shake my hand good-bye.

Lucas was laughing so hard he could barely steer the car. "You shoulda seen the look on your face when he showed you that hip replacement!"

I was as grateful to Lucas for making light of the moment as I was that he had gotten me out of there as quickly as he had. Soon I was smiling, too, and my queasiness subsided. Once we'd settled down, I asked, "How much of your business is cremations?"

"It used to be fifty-fifty, but cremations are on the rise, what with the recession and graveyards runnin' out of space."

"What does the recession have to do with it?"

"Cremation is only about $1,500 compared to a $7,000 casket funeral."

"No way!" I'd always balked at the idea of being burned, but it was hard to argue with those prices.

He nodded, but his expression was grim.

"I take it you're not a fan of cremation?"

"Cremation is just picking up bodies, refrigerating them, dropping them off at the crematorium, and picking up their ashes later." He sniffed. "You might as well be a chauffeur."

"So you appreciate the theater of the open-casket funeral?"

"When someone tells me, 'He hasn't looked this good in twenty years,' that makes me happy. I've done them all—kids, murder victims, suicides. I even embalmed my grandmother."

"Ugh! Really?" I physically recoiled in my seat. I wasn't sure if it was the idea of draining your own grandmother's blood or seeing her naked that threw me more.

"She gave me permission, you know, *before*." His tone was defensive. "She knew how much I enjoyed it." His phone rang. "Oh, hi, Terry. . . . You want me to go right now? . . . Okay . . . Bye."

Lucas hung up and reached for the clip-on tie on the dashboard. "We have a pickup."

Outside the local hospice, visitors glanced up and immediately looked down as we made our way, ominously, through the parking lot, the funeral home logo on full display. Lucas pulled around back and reversed until the van was practically flush against the rear exit. Using the rearview mirror, he fastened the tie beneath his throat.

"You'll have to wait in the car," he said apologetically. "Might be family in there who wouldn't take kindly to having someone standing around just watching."

While I waited I pulled out the *Coping with Death* pamphlet I'd picked up at the funeral home. The introduction explained that at one time, death had been an integral part of family life. People died at home surrounded by loved ones. Adults and children experienced death together, mourned together, comforted one another. Today death is lonelier. Most people die in hospitals and nursing homes. Their loved ones have less opportunity to be with them and often miss sharing their last moments of life. The living have become isolated from the dying; consequently, death has turned into something mysterious, something to fear.

As I read this, I thought about my grandparents; all four of them died within a year and a half of one another when I was in college. Every phone call to my dorm meant a last-minute flight to Texas and a return flight, sometimes still wearing my black dress. At the funeral, I would stare at the casket, willing the knowledge of their death to sink in. There was something about the experience that just didn't feel real.

Lucas returned ten minutes later with the stretcher, covered in a green felt blanket embroidered with the name of the funeral home. On top of the blanket sat a pair of neatly folded eyeglasses.

When my paternal grandmother died, everyone gathered at her house for the funeral reception. At one point I passed by my grandfather's office and saw my grandmother's empty wheelchair parked in the middle of the room. Somehow the emptiness of that wheelchair was more moving than the utter fullness of her casket. I felt a dull ache in my throat thinking of that wheelchair now as I stared as these forlorn-looking glasses.

"Hospices are always easier," Lucas said cheerfully as we pulled away from the building. "It's hard when they die at home and the family watches them leave their house for the last time." When we got to the stop sign, he stepped on the brakes a little too aggressively and, once again, the stretcher crashed into the back of his seat.

He turned and gave the body an admonishing look. "Oh, lighten up."

⌐

After leaving work, I grabbed some fast food, which I wouldn't let myself eat until I'd washed my hands three times. Back at the motel, I took a shower to rinse off any ashes that might've settled on me at the crematorium and dried myself on the small stiff towels. I changed into a pair of Matt's boxers and a nightshirt and climbed into bed with a couple of Eleanor books. I wanted to learn how death affected her life. Franklin passed away suddenly at age sixty-three, but his doctors had been worried about his health for some time. His blood pressure registered 240/130 while he was campaigning for his fourth term in office. A cardiologist was called in and forced Franklin to cut his cigarette habit from twenty or thirty a day to five or six. But the damage was already done. On April 12, 1945, Eleanor was called away from a benefit and summoned back to the White House.

"I knew in my heart that something dreadful had happened," she later said. They told her Franklin had died of a cerebral hemorrhage at his winter home in Warm Springs, Georgia. The First Lady flew to Georgia immediately. When she arrived at the cottage, two of Franklin's cousins—Laura Delano and Margaret Suckley, who'd been vacationing with him—sat her on the couch in the den and told her the story. Franklin had been in a good mood, laughing it up with his visitors while posing for a portrait.

"I have a terrific pain," he said suddenly, his hand flying to the back of his head. Then he collapsed and never regained consciousness.

When Eleanor asked about the portrait, they admitted it was commissioned by Lucy Mercer, the woman Franklin had had an affair with thirty years prior. Now a widow, Lucy had planned to give the painting to her daughter as a gift. Lucy had been Franklin's guest for the past few days and was in the room when he died. Eleanor calmly asked if Franklin and Lucy had seen each other before this final visit. Laura confessed that Lucy had been a guest at Warm Springs several times. When Eleanor was traveling, Lucy often joined Franklin for dinner parties at the White House. It was an open secret that Eleanor wasn't in on. Worse, she learned that her own daughter, Anna, had arranged many of their rendezvous.

"Wow," I murmured. "That's some *Dynasty* shit."

Yet Alexis Carrington would've been disappointed in Eleanor's reaction. She sat there silently for a few moments, processing the information. Then she rose from the couch, walked into the bedroom where her husband's body lay, and shut the door. When she came out a few minutes later, she was still dry eyed and perfectly composed. She was emotional, I suspect, but after a lifetime of practice, she'd learned to manage her feelings when she had to. She never commented publicly about her husband's dalliances, but she alluded to it in one of her autobiographies.

"Men and women who live together through long years get to know one another's failings," she wrote. "He might have been happier with a wife who was completely uncritical. That I was never able to be, and he had to find it in other people."

When I walked into the funeral home the next morning, the phone was ringing manically and Terry was holed up in his office taking calls.

Lucas yawned. "We had three deaths last night." He was stretched out rather unsuccessfully on a love seat in the dining room.

"You did all the pickups yourself?"

"That's nothing." Lucas took off his glasses and rubbed his eyes and I saw that he was a pretty cute guy. "One time we had seventeen deaths in one weekend. We lined them up three deep in the garage."

When Terry was overloaded, he occasionally had other funeral directors come in and freelance. Which is how I ended up in the preparation room with Sean, a funeral director from Columbus, about to witness my first embalming. Embalming made me more uneasy than I'd been before seeing the bones yesterday. Bones were anonymous. You never knew who they'd been attached to. But there was something primal, almost satanic, about draining the life force out of someone. So I was relieved that Sean was the human equivalent of Winnie-the-Pooh—blond, round, and gentle of tone (but, thankfully, wearing pants).

On the stainless steel embalming table lay the naked body of a woman in her seventies. Her skin had taken on a yellow cast. Her gnarled yellowish-green toenails extended far beyond the ends of her toes.

"The Mennonites are hardy folk," Sean said, and I detected a faint Irish accent. "They don't care much about keeping up their pedicures. Around here we call them 'the plain people' on account of their simple dress and because they work in the fields."

Next he opened her eyelid and placed what looked like a spiky plastic contact lens on her eyeball.

"To keep her eyes closed," he explained. "If the eyelids start to open, they'll catch on the barbs."

As I watched him sew the insides of her mouth shut, I thought of that old Dennis Miller joke. "This has to be the easiest job in the world. Surgery on dead people. What's the worst thing that could happen? If everything went wrong, maybe you'd get a pulse." It was surreal seeing an injury inflicted on someone who felt no pain.

He grabbed a few bottles of formaldehyde from the closet and set them down next to the embalming machine.

"Decomposition requires warmth and moisture, so preserving a body means drying it out as much as possible. That's where embalming comes in."

The smell of formaldehyde slipped into the air. Not the invasive, eye-burning smell of chemistry class, but a whiff, like talking on my grandmother's Bakelite rotary phone from the 1950s. As he poured the mixture into the large clear cylinder of the embalming machine, he brought to mind a witch bending over a cauldron.

"It looks like blood," I noted uneasily.

"It's dyed that color on purpose. To bring back that rosy hue to the skin."

Using a small blade, he made a four-inch incision near her collarbone and inserted a metal tube into her carotid artery. Then he made another incision on the opposite side of her neck and inserted a metal tube into the jugular vein. With a series of clicks, the machine began pumping the embalming fluid into the body. From the carotid artery, it would travel through the circulatory system, pushing out the blood, which would flow out of the jugular vein and onto the table. Lining the perimeter of the table was a gutter which would catch the runoff and whisk it away into a funnel-shaped receptacle near the woman's feet and, eventually, the sewer system.

"Luckily, Mennonites don't believe in autopsies," Sean said. "An autopsied body takes three to six hours to embalm because all of the organs have been removed. We have to go in and track down all the different arteries and embalm the arms and legs separately."

As the embalming fluid snaked through the woman's circulatory system, her complexion pinked up, just as Sean anticipated.

"Is that a C-section scar?" I asked, pointing to her abdomen.

He frowned. "That's unusual. Amish and Mennonite women almost always have their babies at home."

"I wonder what went wrong that caused her to go to the hospital," I said, talking more to myself than to Sean. The scar reminded me that there was once life here.

"So do you feel like you're at peace with death?" I asked.

"I thought that I was. As a funeral director, you understand that it's a natural part of life. Yet when my mother died, I was completely devastated." There was a faraway look in his eyes. "She always sat in the same seat at the dinner table, and there was a lightbulb over her chair. It was really uncanny because the lightbulb never had to be changed. It stayed lit for years. Then the day she died the lightbulb burnt out. When I saw that, I just fell apart." He shook his head and smiled sadly. "So yes and no is the answer to your question."

Not knowing what to say, I said nothing. I felt incredibly guilty for stirring up such a painful memory. We stood in silence for a long time after that. An hour and a half and four gallons of embalming fluid later, Sean poked the woman's arm with his finger and gave an appraising nod.

"She's firming up real nice." He turned off the machine. While he sutured her neck, I gingerly touched her arm. She felt, as I'd anticipated, cold and stiff. *This wasn't so bad,* I thought, exhaling with relief.

"Hey, what's that?" I asked.

Sean held up an ominous-looking tool attached to a suction hose. It was a hollow metal rod about two feet in length, but sharp on one end like a spear. "This is a trocar. It's time to aspirate."

Before I could respond, he plunged the spear into the woman's abdomen. Startled, I jumped back a little and watched in horror as he proceeded to empty her of any blood left behind. This was the moment I started thinking seriously about cremation.

A few minutes later he extracted the trocar and popped a plastic plug into the hole in her abdomen. Then he pulled out a fat tub of moisturizer and a hardware-store paintbrush and dabbed the cream on her face and hands to keep her skin from drying out.

"I think she came out pretty good myself," he said with obvious pride. "You came out pretty good too. You could be a funeral director."

I smiled weakly. That night Chris called and relayed a story about something funny that had happened at work. I was grateful for a distraction from the events of the day. Just before we hung up, I asked, "What do you think happens when you die?"

He was quiet for a second. "I have a very vivid memory of being maybe nine years old and sitting in the back of my parents' car as we were driving somewhere in rural Maine," he said. "It was a time in my childhood that I was really afraid of dying. I was looking out the window at the trees as they went by. Since it was so dark, you could only see them in front of and alongside the car. But as they slipped past the headlights, right beside my window, they would just wink out and I couldn't see them anymore. And I remember thinking that death was like that. Suddenly you were gone and there was nothing."

The next morning I woke up feeling better, knowing that the worst was over. When I arrived at the funeral home, I heard voices coming from the prep room and slipped inside. The room was full of people. Not just the three dead bodies lain out on the stainless steel tables, but also Sean, Lucas, and a handsome Italian-looking guy who was built like a fire hydrant.

"Noelle, this is my partner, Antonio," Sean said. "We operate a funeral home together in Columbus."

"I'm not your *partner*," Antonio countered. "I'm not gay like you."

"Gay, am I?" Sean guffawed. "Gay with a wife and seven children?"

"Overcompensating," Antonio said with a smirk.

"How's your divorce going, by the way?" Sean asked. Then he turned and gave me a wink.

Antonio's smirk disappeared. "Lorraine and I are in *counseling*," he said huffily. "That's not the same as divorce."

While they quibbled I checked out the body Sean had been working on when I got in. He was extremely thin with brown hair and a matching beard framing his haggard face.

"Um, is it just me or does this guy look exactly like Abraham Lincoln?" I asked.

"Totally!" Lucas called out from across the room. "That's what I said when he came in."

"Have all of these bodies been embalmed already?" I asked Sean.

"They have. Now we're preparing them. Makeup and such."

"And such" turned out to include plunging a disturbingly long needle into the man's eye socket and injecting it with pink gel.

"After people die, their face changes immediately," Sean explained. "It sinks in. We had to build the tissue back out." He inserted the needle into the side of the man's face and his hollowed cheekbones rose up like baking bread. Sure enough, he looked almost . . . alive at the end. My mind went back to the stuffed birds at Springwood that Franklin shot as a boy, which had been posed to look as though they were flying. It was slightly unsettling but I felt okay.

Justin appeared behind us. "It's open casket. Here's what he asked to be dressed in during the service." He handed me a pair of folded khakis and an orange and brown shirt.

"Is this a . . . Cleveland Browns jersey?" I asked in a disbelieving voice.

Sean shook his head. "It's bizarre what people want to be dressed in."

I helped Sean awkwardly maneuver Abe's cold rigid limbs into the clothes. "Do you know where he stands on tucked in versus out?" I asked. "Because it really sets the tone for the whole outfit."

He left the decision to me and turned to his next customer, a Hispanic guy in his late forties. He pulled off the sheet, revealing the man's naked body. "Jaysus, Mary, and Joseph!" he exclaimed.

I whirled around and was faced with the biggest penis I'd ever seen.

Antonio came over and squinted at it. "Was this guy related to a horse?" he asked. "Can you imagine how big that thing was when it was standing at attention?"

"He's got *my* attention," Sean said.

"See, I knew you were gay!" Antonio crowed, and Sean rolled his eyes.

"What did he die of?" I asked.

"Leukemia," Lucas said. "Picked him up from the hospice yesterday morning. Name's Ortiz."

I'd overheard Terry talking about him yesterday. There had been a problem with the death certificate. They hadn't been able to fill it out because no one knew Ortiz's birthday. He had no family or friends, and the nurses were trying to scrounge money together for a funeral service. How sad, no one knowing your birthday, I'd thought. A few years ago, my friend Rob's parents died, and though he was in his forties, he felt like an orphan. It was strange, he said, that there was no one left in the world who'd known him all his life.

"Lucas, your girl looks like shit!" Antonio shouted from the other side of the room. "What the fuck did you do to her?"

Sean and I abandoned Ortiz to see what he was yelling about. We crowded around the table. Sean let out a low whistle.

The "girl" Antonio had been referring to was a sixtysomething African American woman weighing around four hundred pounds. The trademark Y incision of an autopsy stretched from her shoulders to her abdomen, the oversized stitching reminiscent of a baseball. What really stood out was how bloated she was—not her body but her skin. Water was oozing out of her very pores. Large wobbly blisters were bubbling up all over her body.

"That, my dear, is called edema," Sean told me. "When you're in the hospital and there's nothing they can do for you, they just pump you full of fluids to make you comfortable. And all that water has to go somewhere."

"How do you make it stop?" I asked.

"Usually embalming takes care of it," Sean answered.

"Are you sure you embalmed her, Lucas?" Antonio asked. "Because if so, you did a crappy job."

"She was autopsied!" Lucas said defensively. "You know that makes 'em harder to drain."

"That's no excuse," Antonio tut-tutted.

"I even used Purple Jesus on her!"

"What's that?" I murmured to Sean.

"You use different embalming fluids based on the condition of the body. Purple Jesus is one of the strongest. It's used on the most difficult cases, like to flush out the jaundiced tissue of an alcoholic, for instance."

"Why do they call it Purple Jesus?"

"Because if Purple Jesus doesn't work, nothing will save you."

I was wondering who was going to save Lucas from Antonio, who was still lecturing. "Next time just call me when you get a tough case like this. Because I don't like having to come down and clean up other people's mistakes." He furrowed his brow and leaned in closer to the body. "Her tongue is coming out, for chrissakes! Did you even bother sewing the mandibles shut?"

Lucas stammered, "Uh, well, I—"

"Never mind," Antonio cut him off. "We'll just superglue the mouth closed and be done with it."

Sean tried to change the subject. "Lucas, how about you show us her burial clothes?"

Lucas returned with a mustard yellow dress on a hanger. "Her family says it was her favorite outfit."

Antonio stared at the outfit incredulously. "When? In 1965? There's no way she'll fit into that. It's half her size!"

"What if we just cut the clothes down the back and tuck them in around her?" Sean suggested.

"You can do that?" I asked.

"Oh, that's a classic," Sean said. "Funeral directors do it all the time."

"Well, we can't dress her or put her in the casket now," Antonio pointed out. "We're going to have to wait until right before the funeral. Otherwise she'll just leak all over everything."

He pawed through his supplies bag and emerged with something clear and plastic that looked like a cross between a jumpsuit and a body bag. "Between now and the funeral, let's try to dry her out as much as possible."

A few minutes later the three of them had her legs hoisted up in the air toward the ceiling. I stared wide-eyed as they attempted stuffing her into the suit, which was too small for her.

"Terry knows I charge by the pound, right?" Antonio grumbled.

"He's only joking," Sean assured me. "If we did that, someone Lucas's height would be, like, one-fourth the normal cost."

Antonio hooted, his good mood restored.

A dull click emanated from somewhere in her body. "Uh-oh," Lucas squeaked. "I think I just broke her kneecap."

Yep, definitely cremation for me, I thought.

Once she was inside the bag, Antonio sprinkled her body with a special blue powder designed to soak up liquids, then turned to load Abe Lincoln onto a gurney. "Noelle, can you help me get him into his casket?"

"Why am I doing all the cleanup?" Lucas whined. "I got the short end of the stick today."

"You got the short end of the stick in life, too!" Sean cracked.

Antonio cackled and added, "Listen, Lucas, it's like I tell my wife—shut your hole and know your role!" He casts a sideways glance at me. "Are you married?"

"No, I'm not," I said, caught off-guard.

"Good! Don't get married. Trust the undertaker—life's too short."

Next to the preparation room there was a smaller garage full of caskets. This was the holding area. Like the green room where dead people could hang out after they'd been through hair, makeup, and wardrobe, but before it was time for their show.

As we wheeled Abe into the garage, Lucas called after us, "Remember, it's only a temporary coffin. After the service tomorrow, he's going to be cremated."

"You can *rent* coffins?" I asked in disbelief.

"Oh yeah. Though, really, it's more of a sublet."

I was nervous to pick up the body and afraid that I'd drop him. I braced myself, expecting to feel the heaviness of a man, but Abe was surprisingly easy to lift. There seemed to be almost no mass. It was as if most of the weight was contained in the soul. I felt a sudden rush of pride as I helped maneuver him, honored to be a part of such an intimate ritual.

We placed Abe carefully into a casket, which was, appropriately, the color of a shiny new penny. Antonio arranged the man's hands on his abdomen, left hand clasped over right.

"You forgot this!" Lucas ran in and placed a plushy orange Cleveland Browns football in the crook of his arm. The three of us stared into the casket for a few moments. There was something very dear about this display.

"Well," I said, "now I know what Abe Lincoln would look like reimagined as a Browns fan."

Lucas shook his head in amazement. "It's not what I'd wear to the hereafter."

"Me neither," Antonio said. "What if God is a Steelers fan?"

For her journey into the afterlife, the African American woman had pre-selected an extra large white casket with copper piping. The Cadillac of caskets, really. When I arrived for her funeral the following day, I was speechless. She looked fabulous. Antonio had worked his magic. You'd never know her clothes were cut up the back. He'd even put a veil over the casket to keep people from touching her. Genius. Terry and another funeral attendant stood on either side of the casket as sobbing, moaning family members came up to pay their last respects.

"We've had people flinging themselves into the coffin before," Lucas whispered as we stood in the back, handing out programs to latecomers. "Now we have people there to hold it to make sure it don't get knocked over and spill the body onto the floor."

A half hour later we said good-bye to Abe Lincoln in all his football jerseyed glory. I thought he'd be the most casually dressed person at his funeral, but I was mistaken. Outfits seemed to have been selected with no other criteria in mind besides what best showcased their tattoos. One of Abe's daughters arrived in a halter top and black denim hot pants just long enough to cover her butt. Her brother, to his credit, had worn his best Nike swoosh pants. The eulogy was delivered by a chaplain named Biff. Just as I was admiring his button-down shirt, he turned around to reveal a massive dragon embroidered on the back.

After the service, Lucas took Abe Lincoln to the crematorium, and I headed home for the day. On the way I passed by a horse farm with rolling hills. As I watched the horses flicking their tails delightfully while they grazed, I felt strangely content. If I'd had my own tail, I'd have been flicking it. I was almost too ashamed to admit it, but I'd actually been enjoying myself. It struck me how much I'd miss the place and the people I'd met here. But underneath all this was an urgency. Tomorrow was my last day and I had no more insight into my fear of death than when I'd arrived.

When I got back to my hotel room, I shut the blackout curtains so no one walking by could see in. Then I took the pillows off my bed and settled down on the greasy bedspread, my back straight against the headboard. Dr. Bob told me that meditating didn't just have to be about observing your breath and detaching from your thoughts. You could also meditate to receive insights on things you don't understand. I began by stating my intention: *"I want to better understand my fear of death."*

Then I closed my eyes and let everything that came to my mind rise up, without trying to block any thoughts. Surprisingly, what came up wasn't a thought, but a story from a book I'd bought at the beginning of the project called *Courage: The Joy of Living Dangerously*. I'd almost forgotten about that book. It contained an Indian fable about a powerful emperor who died and went to heaven. According to the legend, every thousand years when a very important emperor died, he was given the honor of engraving his name on the highest mountain in heaven, which was made of solid gold. So the emperor hiked all the way to the top but was baffled to find there was no space for his signature. The whole mountain was engraved with names of past emperors! The emperor was crushed, having finally understood his insignificant place in eternity. Heaven's gatekeeper, who was watching with great amusement, suggested the emperor erase one of the other names to make room for his own.

"What is the point?" the emperor replied bitterly. "Someday somebody will come and erase it."

That was what it meant to face death, I thought. Having to face your own impermanence. Fear of death was the fear of being nothing. The fear of being so easily erased, your presence on earth replaced by someone else. Eventually, everyone who remembered you would die and you would be forgotten. It would be as if you were never there at all. It was a terrifying thought. I felt the clouds part in my mind slightly, some larger understanding beginning to shine through. It wasn't dead bodies or the physicality of death that I was afraid of, not really.

"Accept uncertainty," Dr. Bob was always telling me. Death was the biggest uncertainty in life. You couldn't prepare for it. You never knew when it would come for you. When it did, you were stripped of everything familiar. You couldn't take anything with you. You had to go alone. All fears were a process of letting go, I realized, and death was the ultimate release. You accepted that the world would go on without you.

The next day I rode with Terry in the hearse to deliver Abe Lincoln's ashes to his family and the Mennonite woman's body to her funeral service at the cemetery, my last jobs before heading home. At stoplights the cars next to us hung back, curious drivers angling for a glimpse of the coffin through our curtained windows in the backseat. I tried to imagine what Franklin's cross-country funeral procession must've been like for Eleanor, having to grapple with so much at once— her husband's death and infidelity, her daughter's betrayal, staying strong for the public. She'd stayed the night in Warm Springs and the next morning boarded a train that carried Franklin's body back to Washington, D.C. She kept the window shade up the whole way and looked out at the thousands of weeping Americans who gathered along the route in tribute. She wore one piece of jewelry at the White House funeral service, a gold fleur-de-lis pin that had been a wedding present from Franklin. When she returned to her apartment in New York, a clump of reporters was stationed on her doorstep. "The story is over," she told them.

We pulled into the Mennonite cemetery and drove the hearse toward a group of men in straw hats and women in bonnets and navy prairie dresses. Like the Amish, they wanted to run the funeral themselves, so Terry and I stayed back at the hearse while they performed the service. When the service was over, they headed to the woman's house, where the entire community would gather for lunch.

Next we delivered Abe Lincoln's ashes to his family. When I handed the urn off to Abe's son, he revealed that after yesterday's service, they celebrated by playing the video game Rock Band for six hours with each family member playing a different musical instrument. Now they were strategizing about how to carry out Abe's final wish to have his ashes buried in Cleveland Browns stadium, which was illegal. So far the plan was to walk to the front row during a game with the ashes disguised in a coffee cup, pour the cremains over the side, and hope they were faster than the security guards.

That night, as the train pulled me back to New York, I stared out at the trees flashing by. When I thought about the people I'd seen crying at the funerals that day, I knew Sean was right—this was a fear you couldn't practice for. No matter how many loved ones you'd already lost, it would always be devastating. We're human beings. Even more powerful than our instinct to stay alive is our instinct to love. It's why people run back into a burning house to rescue their family members. It's why mothers throw themselves in front of their children in the face of danger.

I wouldn't say I no longer had a fear of death, but I'd made peace with it. Death had been demystified for me. I wasn't sure I'd want to be completely rid of that fear. Fear of death could go one of two ways: it

could force you to live in the present, where you had a greater apprecia-
tion for the people and things around you, aware of the fragility of life;
or it could force you to live in the future, always worrying about when
death was coming, but then you weren't really living.

In the last week I'd met the most interesting people of my life, not
all of them alive. Some of us spent our lives working in the fields and
had quiet deaths. Others went out with style, wearing our favorite outfit even when it was too small for us. We were all so different. Death
was the one thing we had in common. There was something incredibly
lovely about that.

Chapter Fifteen

◡

We are constantly advancing, like explorers, into
the unknown, which makes life an adventure
all the way. How interminable and dull that
journey would be if it were on a straight road
over a flat plain, if we could see ahead the whole
distance, without surprises, without the salt of the
unexpected, without challenge.

—ELEANOR ROOSEVELT

The truth was that I wasn't ready to climb Mount Kilimanjaro. I
knew Dr. Bob would say, "You can't wait until you feel 'ready' to
take a risk. You'll never feel truly prepared," but I truly was not
prepared. As in, I didn't have any clothing or gear yet. Mentally and
emotionally, I didn't know what to expect. I was physically ready, at
least. For the last two months I'd hit the gym three days a week, climbing the Stairmaster and squatting into unbecoming positions with a
weighted barbell balanced on my shoulders. But that was about all I
had going for me.

In my defense, Kilimanjaro was a hard mountain to prepare for.
It was a mountain of extremes comprising five distinct climate zones:
rain forest, heather, moorlands, alpine desert, and ice cap. Tempera-

tures ranged from the eighties in the rain forest to negative fifteen degrees at the summit. A typical hike took four and a half days to reach the summit, and only a day and a half to descend. Kilimanjaro inspired extreme behavior in others. In 2001, an Italian man named Bruno Brunod (seriously) ran to the top in a record five hours and thirty-eight minutes. Wim "Iceman" Hof did it in two days, bare chested, wearing only shorts. Douglas Adams, author of *Hitchhiker's Guide to the Galaxy*, climbed to the summit wearing an eight-foot rubber rhinoceros costume. I'd stick to hiking gear, thanks. Though, frankly, you'd have had a better chance of finding a rhinoceros costume in my closet.

Thank God for Becca, who came to my rescue on the clothing front. While mixing milkshakes one day, I told her I wouldn't be able to volunteer at the hospital for the next two weeks because I was climbing Kilimanjaro. "No way!" she squealed over the blender. "I climbed Kilimanjaro three years ago!"

"Really?" Becca was so girlish that I would have never thought to ask if she was a mountain climber.

She stopped the blender and poured the milkshake into a cup. "Yes, I wanted to see the glaciers before they disappear."

As we rolled our cart of milkshakes down the hall for delivery, she said quietly but firmly, "You *will* get diarrhea, by the way. It's just a question of when."

Most of her hiking stuff was at her parents' house in Maryland, she said, "but I can bring what I have to your apartment a few days before you leave. And if you want, I can look over your clothes and equipment to see if there's anything else I think you might need."

"Like someone to carry me up the mountain?"

"Oh please, you'll be fine!" she said cheerfully.

The tour company I'd signed up with had a checklist on its website detailing what I'd need for the hike. I printed it out and crossed off the items I'd be borrowing from Becca. The rest I'd have to buy. I took my list to REI sporting and camping goods store, where there was nothing sporting or good about the merchandise total. It had to be split up on four different credit cards. The hiking boots alone were the ugliest $200 I'd ever spent. Standing there in front of the cash register, it hit home that I'd officially bankrupted myself with this project. I had a moment of reckoning with my bank account. (*"Oh my God, you're EMPTY? HOW COULD YOU?"*)

A few days before I left, Becca brought over her gear. It was a thirty-minute subway ride and ten-minute walk to my apartment from hers, but when I'd offered to pick it up she'd waved me off, saying, "I need to check out what you've bought and make sure you haven't forgotten anything." Looking at the heap of stuff she'd toted, my heart swelled a bit, the way it does when someone you don't know very well makes a gesture of generosity that exceeds the nature of your relationship. In that instant I felt us shift from acquaintances to friends.

"Thank you so much for doing this," I said. "I really appreciate it."

She examined the clothing and hiking equipment laid out on my bed. "The only time it'll be warm enough to wear shorts is at the very beginning and end of the hike," she said, tossing all but one pair into the Don't Need pile. She suggested bringing extra batteries. Cold weather drained them quickly, which could result in a particularly disappointing moment at the summit when one realized his or her camera was dead and the nearest convenience store was twenty thousand feet down.

"When you get to the colder temperatures, keep the batteries as close to your body as you can, even when you're asleep. Your electronics, too. Before going to sleep, stuff them in the bottom of your sleeping bag where it's warmest."

A half hour later I walked her to my door where we hugged goodbye. "Remember," she said, "the only thing that gets you up that mountain is sheer iron will. On the last part of the summit climb you're actually on all fours crawling to the top—partially because of the steep angle but also because you're so exhausted."

Noting my worried expression, she added: "Just remember, you can always take another step."

The night before I left, I stopped by Jessica's apartment to drop off my keys so she could feed my parakeets while I was away.

"So how are you feeling about it?" she asked.

"I've never traveled so far from home before," I said nervously. "Especially not alone and to a third world country."

She pulled me into a hug. "Listen, I love you. You're about to go on an incredible journey. You'll learn something on this trip that you can only learn by confronting this mountain and what it represents." Jessica had become gentler in the past few months. Surprising us all, she'd gotten heavily into yoga and was even considering signing up for a spiritual retreat in the Berkshire Mountains. She pulled back and held me at arm's length. "Namaste, bitch," she said. "Oh, and bring me back an orphan baby. Obviously."

The first day we'd hike through rain forest to 9,000 feet, where we'd spend the night in the Mandara Huts. The second day we'd cross the heather and open moorlands to the Horombo Huts at 12,000 feet. The third day, our acclimation day, we'd be hanging out at 12,000 feet, giving our bodies time to adjust to the thinning oxygen. The fourth day we'd climb over alpine desert to Kibo Hut, which, at 15,000 feet, was the last campground before the summit. That night we'd go to bed early, awaken at midnight, and hike six hours to the ice cap summit at 19,340 feet. Then we would turn around and hike back down to the Horombo Huts, spend the night at 12,000 feet, and hike the rest of the way down the next day. Altogether we'd walk more than fifty miles.

But first I had to get through customs. As I stood in line with the other grungy hikers, my backpack felt aggressively oversized, my gear too new. I scanned the airport crowd until my gaze settled on a portly African man holding up a sign with my name on it. He'd be driving me to the city of Arusha, where I'd stay in a hotel tonight and tomorrow recovering from jet lag before hitting the mountain. When I said that I'd be staying in a hotel, I meant that I'd be *staying* in a hotel. "Sightseeing in Arusha is discouraged" was a common refrain on Kilimanjaro websites. On the city's Wikipedia page, this line stood out: *Increasingly, tourists are being held up at machete point, even during the day.* ("Ohmigod, that's so authentic!" Jessica had said.)

It was only 7:30 P.M. but felt much later because there were no streetlights. There were other people on my shuttle staying in Arusha, most of them bound for safaris. The scene was the same at every hotel. Each had a gate out front guarded by a formidable canine that greeted us by straining at the leash, rising on its hind legs, snarling. Holding the leash was a young man in a camouflage uniform. Another camo-garbed attendant with a long-barrel gun slung over his shoulder stepped forward to check the driver's affiliation, opened the gate, and immediately closed it behind us. The presence of the guards was both comforting and alarming. I felt safer knowing they were out there, but why were they necessary? My hotel had an open-air lobby with a dingy tile floor. It was completely empty except for the receptionist, who was trying to program the Stevie Wonder song "I Just Called to Say I Love You" as her cell phone ringtone. She paused long enough to hand me a cartoonishly big hotel key. The bellhop was a teenage boy swathed in a plaid blanketlike garment that extended over his head. In place of shoes, he wore pieces of rubber car tires on his feet. He led me down a winding concrete path lined with tall plants to a room on the first floor. The room was bare bones with the same tile floor as the lobby and stone walls. Because I hailed from a first world country, the mosquito net over the bed seemed romantic, instead of reminding me of malaria. But the single lightbulb hanging by a cable from a hole in the ceiling had a nooselike quality that had me swallowing nervously.

I'd been awake for thirty-two hours. In that amount of time, I'd packed, traveled from New York to New Jersey to Amsterdam to Tanzania. I'd flown seventeen hours, watched eight movies, and endured a five-hour layover. Still, I wasn't tired. So I opened my hiking backpack and pulled out two rectangles of white poster board I'd brought from

New York. Becca had given me the idea. "You should make signs that you can hold up in the pictures you take at the summit," she'd told me the last time I'd seen her at the hospital. "They make great Christmas presents." I'd meant to make them before I'd left New York but packing had taken longer than I'd expected, so in the end I'd just thrown the poster board and two black markers into my backpack. Now I was grateful to have something to do here in this room with no telephone or television or even electrical outlets. Under the harsh light of the naked bulb, I spent an hour making block letters and coloring them in with black marker. On each side I wrote a different message:

HI MOM!

I ♥ DAD

I ♥ MATT

(and just for fun) I'M HIGH

The ink dried up toward the end and I hoped my dad wouldn't notice that the letters on his sign grew progressively lighter from left to right. At 11:30 P.M. I opened my sleeping pill bottle for the last time this week. Over the past five months I'd gradually stepped down to a half pill only to fall off the wagon. After a recent rash of restless nights, I was back up to three pills. As I popped them into my mouth, I felt both guilty that I was still taking them and terrified that I was about to give them up. Tomorrow would be the first time in ten years that I'd gone to bed without some form of sleep aid. When the familiar soupy feeling took over, I climbed onto the mattress and arranged the net around me, making sure there were no unauthorized openings. Somewhere around five in the morning I woke to the beautiful soulful drone of the Muslim call to prayer being broadcast over a loudspeaker somewhere nearby.

My head guide, the man who would be leading me up the mountain, was named Dismas. A thirty-six-year-old native Tanzanian, he had been to the top more than three hundred times.

"Miss Noelley, your name like Christmas. Mine too!" He grinned as I climbed into the van that would ferry us to the base of the mountain. Everything he said sounded exotic because of his lilting Swahili accent, even when he opened up the conversation by saying, "Deed you know dat dee King of Pop has passed on? I see on CNN thees morning."

I'd be hiking and bunking with an older French-Canadian couple, Marie and Henri. The first thing I noticed about them was that they had nearly identical bodies, five feet seven, sturdy but trim. They were outfitted in T-shirts and hiking pants made of synthetic, sweat-wicking textiles. Henri's head could have been lifted off the neck of actor David Strathairn. Marie had the clean blank face of a woman who rarely wore makeup. Her thick brown hair was cut bluntly above her shoulders and jutted out several inches from her head. They'd returned yesterday from nearby Mount Meru, a fourteen-thousand-foot mountain they'd hiked to practice for Kilimanjaro.

"To better acclimate ourselves to the altitude," Marie said brightly.

Marie was a nurse who used to work in an oncology ward. She and I talked about her job and my volunteer work as we drove, cars fffftttttting by dangerously close on the two-way street. Henri stared out the window quietly. It was a blur of trees, hills, and the occasional stream, interrupted by boxy one-story buildings and concrete general stores with soft drink and beer logos painted on the sides. We passed a leafy coffee plantation where women stood barefoot in the plants rummaging for beans. I marveled at the women walking alongside the road with baskets balanced on their heads but was completely charmed by the little girls trailing after them who also had baskets on their heads; being less steady than their mothers, they kept one hand on the top at all times. Though it was in the eighties, most of the men milling about wore button-downs and jeans or khaki pants. The rest were in native dress, their bodies swaddled from head to toe in brightly colored material, with matching head scarves. Some of them carried walking sticks, which they used to prod herds of goats and cows. There were a number of stray dogs and cats lurking around. Whenever I traveled I found the sight of animals strangely comforting. No matter where you went, they were the only things that looked exactly the same as back home. When our van pulled over, I stuck my head out the window. A uniformed officer at a makeshift roadside guard post was gesturing for us to stop. Our eyes met and he hesitated. Then he changed his mind, silently dismissing us with a wave.

I leaned in to Marie and asked in a low tone, "What was that about?"

"Police checkpoints," she answered. "They randomly pull over cars and search them. If they find anything wrong—and they'll nitpick until they find something wrong—you have to pay them a fine on the spot or they take your car. But when they see white people in the car, they let you go. Bad for tourism."

Three hours after leaving Arusha, we arrived at Marangu Gate, the entrance to Kilimanjaro. As we pulled into the parking lot I was struck by the oddness of a rain forest having a parking lot, as well as a gated entrance. We were greeted by a swarm of young, fit African men. In addition to Dismas, we'd be accompanied up the mountain by an assistant guide and ten porters.

"I feel like a 1930s British colonialist," I whispered uneasily to Marie and Henri as the porters unloaded the van of our personal effects, which they'd carry up and down the mountain for the next six days.

The three of us checked in, writing our name, age, address, and occupation in a book, something we'd have to repeat at every campsite. I signed last and saw that Marie was forty-seven and Henri was fifty-three and a graphic designer. By the time we were done, the porters had already hit the trail with our duffels, sleeping bags, and food for the week. We followed them into the rain forest, breathing in the loamy smells of minerals and chlorophyll. To discourage erosion, logs had been arranged along the path forming a kind of staircase to assist us on our ascent. A group of African kids, maybe seven years old and wearing Crocs in Day-Glo colors, were blocking the trail. Hands outstretched, they repeated "money, money, money." Dismas shooed them off and we continued. Intermittently, porters from other hiking groups approached from behind and we stepped aside to let them pass.

"Jambo!" (Hello!) they called out jubilantly in Swahili, grinning widely. They were a chiropractic nightmare, carrying forty-five-pound bags on the back of their necks, heads tipped forward, for more than six hours a day. A few balanced them on their heads as they walked. For this, they earned an average tip of $5 a day, the same I would give a bellhop in the United States for carrying my bag for three minutes.

After checking to make sure Dismas was out of earshot, Marie said, "They're beasts of burden. Other cultures use camels or mules. Here they use young men."

When I'm in New York, I walk so fast that other pedestrians zoom by, as though I'm on an invisible moving sidewalk. On the mountain, our pace was set by the guide. We were walking wedding-march slow. "Poly-poly" was the motto on Kilimanjaro. It meant "slowly, slowly" in Swahili. Going poly-poly helped stave off altitude sickness and increased the chances that hikers would make it to the summit. No one was more disappointed by "poly-poly" than Henri. He'd set such a brisk pace on Mount Meru that the guides had nicknamed him "Mountain Gazelle," Marie told us with obvious pride. Dismas gestured to the dainty lavender flowers lining the trail. "These flowers are called impatiens."

"Impatiens, huh?" I laughed. "I know how they feel."

"You have luck on your side, Miss Noelley," Dismas said. "The majority of people who make it to the top are old people and women."

"Really? I would've thought young men."

He shook his head. "Their blood is still too hot. They don't go slowly. They rush. Then they have to come down."

When we reached the first rest area, I plopped down next to Marie on a picnic bench. "My ass had better look incredible after this," I told Marie, stretching my legs out in front of me. They were holding up pretty well considering I'd basically spent the last three hours walking up a giant staircase. I guessed my training paid off.

Dismas and the assistant guides divided their time between trying to scare off a mongoose—peeping out of a bush with its little bear face and a mink's long body—and waving away a long-billed crow threatening to swoop in and make off with our sack lunches (a frequent problem on the trail, according to Dismas). When they were not warding off animals, they ate their lunch sitting on boulders about fifteen feet away. There was plenty of room at our picnic table, but when we invited them to join the three of us, they refused, heightening the feeling of segregation that pervaded Kilimanjaro.

In what could only be a miracle from God, there were working toilets at most of the camps and outhouses along the trail. Campers were responsible for their own toilet paper so I'd packed four rolls to last me the entire trip. After lunch I dug a roll out of my backpack and trotted over to the outhouse. I felt a little uncomfortable holding the toilet paper in my hand, such a blatant advertisement of what I was about to do. When I got inside the outhouse, I saw it was just a wooden floor with a hole in the middle. *Well, this is going to be interesting,* I thought. Flies orbited in lazy circles above the hole. As I squatted over it, I wondered if one of them would fly up my vagina and what the protocol would be in such a situation. Thankfully, they fled in terror. My already over-worked quads trembled a bit as I balanced my weight, but the real problem was my vagina, which had always operated less like a hose than a five-nozzle sprinkler. This was fine when you were sitting on a toilet, but now it scattered urine in all directions, sending it racing along my butt cheeks, down the backs of my legs, and into the tops of my hiking boots.

Our first overnight stop was Mandara Hut, a campsite nestled in a misty forest clearing. By the time we arrived in the late afternoon, the guides and porters had been there for hours, wandering the campground's grassy slopes talking on cell phones. One of them had a Bluetooth device clipped to his ear. After settling in, we tramped into the long-tabled dining hall. An abundance of languages could be heard, but the diversity ended there. Except for one Asian group, all the hikers were white. The groups would hike separately but en masse up the mountain, bunking and eating together at the same three campgrounds. The largest group had twenty-three hikers, churchgoers from D.C. who were climbing to raise money for clean water in Liberia. The pastor, who had brought his ten-year-old son, clinked his fork on a glass so ostentatiously that it quieted not only his table but the entire dining hall. Then, in a commanding voice, he recited the predinner prayer. "Lord Jesus, we thank you so much for the bonds that we've formed on this trip and we ask that you guide our conversations at dinner tonight to strengthen our friendships even more. In Christ's name. Amen."

I was scooping lentil soup into my mouth when Marie whispered, "See that guy over there?" I followed her gaze across the dining hall to a man in a wheelchair. "Earlier, I overhead someone say that he's a quadriplegic. His friends are pulling his wheelchair up the mountain with ropes."

As I watched the man being spoon-fed by another hiker, I wondered if he had always been paralyzed. And if not, had they become his friends before the paralysis or after? And which would say more about their character?

After we finished dinner, the three of us repaired to our hut. The huts were actually individual wooden cabins with steeped roofs. Built into the slanted walls were three narrow double-decker beds topped with thin, plastic-covered mats. It was so small that we had to take turns standing in the middle of the room. It was similar to the accommodations in the hull of *The Manatee*, in fact. There were enough cabins at this camp to sleep sixty people in all. Soon a porter brought us boiling water to brush our teeth and quickly retreated to the separate accommodations for porters and guides across the camp. We were not getting up until 7:30 A.M., so I assumed we'd stay up reading or talking for a few hours, but Marie and Henri started readying themselves for bed at 8:00 P.M. and I had no choice but to join them. The dining hall was closed so there was nowhere else I could go.

The huts were unheated so we slept in our fleece hiking pants and sweatshirts. It was the first time I'd had to put on more clothes to go to sleep. I pulled on the blue fleece shirt Jessica had lent me. Knowing that she had worn it made me feel less alone.

We each chose a bottom bunk and unfurled our sleeping bags. They were specially designed for subfreezing temperatures. Unlike regular sleeping bags where your neck and head have to fend for themselves, these came up around the head and shoulders with a small opening for the face. That it was shaped like a pharaoh's coffin—wider up top, tapering toward the feet—was not lost on me. I usually slept on my side but because of the narrow fit, there were only two positions for my arms: straight down at my sides or scrunched up in front of me in the manner of a *Tyrannosaurus*. Trying to fall asleep without sleeping pills while posing as a dinosaur in a padded casket would've been challenging in and of itself. Throw in the fact that my body was still on New York time, where it was 1:00 P.M., and it was not looking good for sleep. For a while I listened to my heartbeat playfully skipping around, trying to adjust to the reduced oxygen. Then I synchronized my breathing with the slow inhales and exhales of Marie and Henri, hoping I could trick my body into thinking it was already asleep. There were no windows in the cabin. It was so dark that sometimes I forgot to blink because I couldn't always tell if my eyes were open or closed. Every now and then I rolled over and switched sides, just to break up the monotony.

Eleanor, despite her high-society upbringing, loved camping. In the summer of 1925, she, Nan Cook, and Marion Dickerman took her sons Johnny, nine, and Franklin Jr., eleven, and two of the sons' friends on a ten-day camping trip to Canada. They piled into Eleanor's seven-passenger Buick with nothing more than two tents, cooking gear, and a first-aid kit. They slept in random farm fields along the St. Lawrence River, stopping in New Hampshire to rent some burros and climb the White Mountains. Eleanor was endlessly game in the face of discomfort. After Franklin contracted polio, he spent a lot of his time sailing off the coast of Florida, hoping the warm waters and climate might have healing properties. Eleanor couldn't sleep in her cabin due to claustrophobia so she slept on deck, though she felt no less vulnerable surrounded by open sea. "When we anchored at night and the wind blew, it all seemed eerie and menacing to me," she later wrote. "Florida's mosquitoes all converged on me . . . I always wound up with enough bites to look like an advanced case of smallpox."

I had no clue how much time had passed. Three minutes? An hour? Occasionally I brought my wrist to my face and pressed a button on my digital watch. The face lit up, emanating a green alien glow. 10:30. 1:04. 1:30. 2:10. 3:33. I was taking an altitude sickness medication called Diamox in an effort to ward off cerebral and pulmonary edema. It sped up the acclimatization process by allowing more oxygen to enter your bloodstream. It was also a diuretic, meaning that I had to pee four times during the course of the night. It was ten degrees outside and the communal bathroom was fifty yards away, so pee breaks required preparation. First I wriggled out of my sleeping bag, struggling not to wake Marie and Henri with the *whisk-whisk* sound of my body rubbing against the nylon. Then I fumbled around in the dark for my heavy North Face coat and hiking boots. Once they'd been successfully zipped and tied, I snapped on my headlamp, basically an elastic headband attached to a flashlight that sits on your forehead. During one particularly harrowing bathroom break, there was a clacking animal noise I'd never heard before. It got closer as I ventured toward the bathroom. Scared, I broke into a run, my headlamp beam bobbing in the darkness. At 4:30 A.M. I checked my watch for a final time and dozed on and off until Dismas woke us three hours later.

~

On the second day the claustrophobic rain forest gave way to rolling hills dotted with shrubby plants and heather trees. The miles-away peak of Kilimanjaro was finally visible. Dismas took a dozen pictures of me with Kili. Later I would discover that my enormous head was blocking the mountain in 95 percent of them. We trekked on and found ourselves in the wide expanse of moorlands. The transition from one zone to another was abrupt, the way the Magic Kingdom in Disney World was divided into different themes: Frontierland, Tomorrowland, Fantasyland, and so on. You could draw a line across the trail where one zone ended and another began. The absurdist landscape of the moorlands could've been created by Dr. Seuss, especially the lobelia trees with slim trunks and bulbous branches that exploded on top in an Afro of fluorescent leaves.

I was lost in my thoughts, marveling over my utter lack of sore muscles, when Marie asked, "Do you think you're going to marry Matt?"

I blinked. "What?" I couldn't believe this question had followed me up a mountain in Africa. Then again, if Michael Jackson's death could make it this far, I guessed anything was possible.

"Since you referred to him as your husband at breakfast, I just assumed . . ."

"I did? No, I didn't." I thought back to our earlier conversation. "I called him my boyfriend."

"You called him your husband."

"You must've said 'husband' just before me and then I said it accidentally. Or maybe it was the altitude talking."

She smiled knowingly. "Well, you said it. Henri and I both heard it." She turned to him and he nodded in confirmation. Henri had been a little pissy that day and when we'd set off this morning, he'd slyly taken the lead and gradually kept increasing our speed until we were approaching a light jog. Dismas had to install the assistant guide in front to slow him down, and Henri had been scowling since. Whenever Marie asked if he'd like one of her energy bars or how he was feeling, he responded in clipped one-word replies. But Marie, unfazed, either didn't notice or didn't care.

We passed a porter who was sitting on the side of the road out of breath. Marie offered to pour half a liter of her water into his empty thermos. He gladly accepted.

"I'm a nurse." She shrugged as we continued on. "I'm used to taking care of people." A few minutes later Henri silently reached across the dusty trail and grabbed Marie's hand. I smiled to myself.

The day before Marie confessed she'd originally thought our guide's name was "Dismal," not Dismas. Now the whole day I kept almost calling him Dismal and had to catch myself. We were wearing pants and T-shirts but pulled on wool long-sleeved tops whenever the fog rolled in, which it did, often, with astounding rapidity. One minute the air was clear; the next I was staring at a wall of white and couldn't see more than twenty feet in front of me.

Horombo was perched next to a cliff. The cabins were A-frame like those at Mandara but painted an ominous coal black, a striking contrast to white clouds that kicked up wispily over the cliff's edge like the foamy waves of Big Sur. At twelve thousand feet, we were above the cloud line. The sunset turned the clouds pink, making them look girlish and slightly ridiculous. I took a walk after dinner, wandering over to the porters' side of the camp. Half of them grinned and greeted me with "jambo," but many treated me as an invader and glared. This was the tension of Kilimanjaro tourism. They knew they needed us for their livelihood and some of them resented us for it. I didn't blame them, frankly.

Every morning we stuffed our sleeping bags into small sacks, which had a strap for easy transport. Getting a fluffy six-foot-long sleeping bag into a two-foot-by-two-foot satchel was always a comedy of errors. I would smoosh one part down, only to have another part boing out. Then I'd shove that part back in and the opposite end would gleefully pop out. So I felt a sense of relief as I rolled my sleeping bag, knowing I wouldn't have to repack it tomorrow morning. We'd be staying here two nights to better acclimate us to the altitude before attempting the summit. When Marie stepped out of our cabin to brush her teeth before bed, Henri and I descended into awkward silence, as we always did when we were left alone. He busied himself with fixing our broken door using my Swiss Army knife, which I'd packed primarily because it had a nail file. I lay down on my bunk and considered the graffiti etched into the wood overhead. Horombo was the halfway point of the climb. Hikers stayed here on the way up, but they also spent a night here on the way down from the summit. Therefore, it could accommodate 120 hikers, twice as many as Mandara and Kibo. All over our cabin walls, hikers had carved for us their impressions of the climb. One anonymous hiker pronounced the hike to the summit "miserable" but added that "the view is worth it." Shana Theobald on 7/20/2007 wrote: "The pain lasts for a little while, but the pride lasts forever. It's mind over matter—you can do it!" Less sentimentally, "JM + BK" instructed me to "go hard, or go home." Marie and Henri had brought along a digital thermometer, and it had become a ritual to take the temperature inside our cabin each night. Tonight it read three degrees Celsius, or a little over thirty-seven degrees Fahrenheit. As Marie and Henri got ready for bed, the hikers in the cabin next to us gabbed about the day, occasionally giggling at some mutual joke.

"Well, I hope those party animals next door quiet down soon," Marie huffed, climbing into her sleeping bag. I checked my watch. It was 7:00 P.M. I felt like a prisoner at bedtime again. I lay awake for six and a half hours. On my third trip to pee, I creaked open our broken cabin door that even Henri couldn't fix and waddled toward the ladies' bathroom. The terrain was more uneven here than at Mandara, so I walked with wide-set legs to better maintain balance. I continued on to the bathroom where someone had been battling (and losing, it appeared) a war with dysentery.

On the way back to the cabin, my left foot slipped on a rock and I fell backward, landing palms and ass down on the ground, looking straight up at the sky. I gasped. Then I leaned back on my elbows and stared. The sky was radiant in its blackness, the stars bright and crisp. I remembered how the New York skyline had glittered that night, almost a year ago, when I'd swung out on the trapeze; how those tiny squares of light had stood out against the sky. But this! This looked like a photo of outer space. Light travels through space in a straight line. It's the atmosphere that causes it to bend and scatters the light, creating the blurry stars and hazy blue-black color that sea-level dwellers think of as the night sky. Up here the atmosphere was thinner, with fewer dust particles and gas molecules to spoil the view. We were closer to the light, yet more enveloped in darkness.

Every morning I waited for Henri and Marie to leave for breakfast. Then I stripped naked in my thirty-seven-degree cabin air and, gritting my teeth, wiped myself down with moist toilettes. This was my "shower." A person has to maintain some civility, even while dry shaving your armpits. Which brings me to another matter of hygiene. Catherine Deneuve once famously remarked, "A thirty-year-old woman must choose between her ass and her face." She had been referring to aging. As you get older, the theory goes, you can either have a thin body or a youthful-looking face—but you can't have both. One month away from age thirty, I was already choosing between my ass and my face. My nose, thinking that it was winter, had been running nonstop since we'd exited the rain forest. My toilet paper supply was dwindling fast. There was no way four rolls would last the entire trip if I kept using it to blow my nose. As an alternative, I turned to the travel face towel I'd bought especially for the trip, renowned for its absorbency. On principle handkerchiefs disgusted me, but so did the idea of wiping my ass with tree leaves.

Just because we were bunking at Horombo two nights didn't mean we got to rest during the day. No, today we were going on a day hike to Zebra Rock, a some fourteen-thousand-foot affair that was supposed to help prepare our lungs for the low oxygen levels we'd encounter on our summit hike. It was a struggle. No matter how slowly I walked, I couldn't catch my breath. I was panting as though I'd just finished a really long sprint, but the feeling never went away. Frustration built until I was angry at everyone and everything that had ever existed since the beginning of time. I was even annoyed at the United States for not using the metric system, because every time Dismas told us how many meters high we were or how many kilometers we'd hiked, I had no idea what he was talking about because I was too fuzzy-headed to calculate the conversion rate.

Mountain climbing is a love affair on fast-forward. The things you once found charming about your partners quickly become the things you loathe. On the first day I'd been delighted in the way Henri pronounced his name, with that uniquely French uptick that sounded like the word should have an exclamation point at the end (On–REE!). Now it was so grating that every time someone said it, my shoulders actually shrunk up and my head twisted slightly to the side. It only underscored the fussiness of his personality, the way he was always messing with his camera, his shirt tucked in (tucked in!) just so. Marie, meanwhile, was relentlessly inquisitive. Always with the questions! But mostly I resented their need to get ahead of everyone else. I thought that was an American thing. And who hiked a fourteen-thousand-foot mountain to *practice* to hike a nineteen-thousand-foot mountain the next day? Freaks.

Now my irrepressible Canadian cohikers were practically running up the trail, and by the afternoon I hated them with the fire of a thousand burning suns. As the space between us widened, their backs grew smaller and smaller until I could crush them between my thumb and forefinger. Dismas hung back with me. "Just poly-poly. That is the best way to reach my main office." He winked. Dismas referred to the peak of Kilimanjaro as his main office. Mount Kili, by the way, was a tease. One moment the peak was visible, standing naked before us. The next she was wrapping herself in clouds, a bashful woman cloaking herself in a bedsheet after a one-night stand.

"Kili is sleeping," Dismas said whenever she was obscured behind clouds. *That must be nice,* I thought grouchily. *At least someone got to sleep in today.*

Three hours after leaving Horombo, Dismas and I arrived at Zebra Rock. It had once been just a black lava cliff, but years of mineral-rich rain had stripped the color so that it was now covered in white stripes. I had to admit, Zebra Rock was striking to behold. Emphasis on the *behold.* While I was admiring the view, I noticed an arrestingly steep hiking trail winding up the mountain next to Zebra Rock. I pulled my sunglasses down the bridge of my nose to stare at it over the tops of the lenses. Then I looked at Dismas.

"Wait, we're not actually climbing up that thing, are we? It's practically vertical. Is that even legal?"

"It's like a narrow stairway to heaven, no?" Dismas answered dreamily.

"It's the road to Hell, Dismas," I said flatly.

"Poly-poly, Noelley. Poly-poly."

"Slowness is not a problem for me, in case you haven't noticed. If I were to go any slower, I'd be standing still." He responded with a grin.

Despite its steepness, it was not as bad as I anticipated. I settled into a groove and my breathing relaxed. Groups of other hikers were gathered at the top, snacking on various provisions. A guy from the church group offered some pieces of dried mango that looked about the way that I felt.

By the time I returned to Horombo, the hikers who had climbed the summit that morning were staggering in looking like the back-up zombies from the "Thriller" video. They were wild eyed and stiff limbed, not to mention completely filthy. During dinner a fourteen-year-old boy rose from the table next to ours and, without a word to his family, walked outside the dining hall. Through the window we saw him double over and vomit three times. A few minutes later we saw a dusty woman being helped to her cabin in the posture of an injured football star, each arm slung over the neck of a porter. Marie and I exchanged worried glances. Henri asked the German couple next to us with telltale sunburned noses whether there had been any snow at the peak. "No, but there was hail on the way down," the man answered.

Hail? No one had said anything about hail! A helmet was the one thing I *didn't* bring. There was no way I was going to make it to the top. How could I climb seven thousand more feet tomorrow? I'd barely made it two thousand feet today. I felt my eyes welling up. I couldn't do this here. I had to get back to my cabin, but first I had to finish my dinner or I wouldn't have enough energy tomorrow. I started cram-ming great forkfuls of pasta into my mouth. I was chewing fast, trying to get it over with as quickly as possible. I bit my tongue hard. I kept going, chasing the pasta with a slice of fried bread. I bit my tongue again, drawing blood this time. I let my fork clatter to my plate. Then, to my horror, I buried my face in my hands and started to cry. Marie and Henri fell silent, the way you do when someone you don't know very well is crying and you're unsure whether to ask what's wrong or let the person be. I whisked the tears off my cheeks with my fingers, composed my facial expression, and stood up.

"I am finished with dinner," I announced and hurried out of the dining hall back to my cabin. A few minutes later, when I had a really good cathartic cry going, there was a knock at the door. Being inter-rupted at the start of your cry is like being interrupted masturbating or accidentally ripping your headphones out of your ears during a good song. I felt a flashing irritation. I opened the door expecting to find

Marie and Henri, but instead Dismas was standing there. They must have said something.

"Miss Noelley, are you sick?" His forehead was furrowed in concern.

"No, I'm not sick."

"No headache? Throwing up?"

"I'm fine, really. Please, I just need to be alone."

"I see you tomorrow morning then." He tipped his cap and walked away. I closed the door and felt my face contorting again, lips pushed out, chin quivering, eyebrows drawn together. I bent over into a gutteral, full-body sob. When I'd signed up for this trip, I'd known that it was important to come alone so I couldn't use my boyfriend or friends as a crutch. But I was suddenly overcome with homesickness. I missed Matt and Jessica and Chris and Con Edison, the gas company that supplied my heat. *I will never take any of you for granted again!* But most of all I missed sleep. I'd been gone for six days, and I couldn't believe it was going to be five more days before I was home again. Then I remembered something Chris had said last week.

"I know it seems like a long time, but let me share a little secret from when I was a rower in college." He'd explained that when he was on the crew team at Yale, he often had to do timed tests on a rowing machine, appropriately named the erg. "They'd be like an hour of the hardest strain of your life, and I'd always tell myself, 'No matter what happens, in an hour this will be over.' Whether I sucked or did great or even if something terrible happened like I tore a muscle, there would be a time in the near future when I wouldn't be doing that activity anymore. It's kind of a wimpy way to think about things, but it works.

"And just remember," he'd added, "if things get tough, eat the sherpa. That's what they're there for."

How bizarre that I'd just finished dinner and Chris, Jessica, Bill, and Matt hadn't eaten lunch yet, that it was summer where they were and spring here. It was like they were living in a parallel universe and I'd time traveled, which I supposed I had. I was living in the future. My dad had gone to China on business a lot when I was a kid. Whenever he'd call, the first thing I'd ask was, "What day is it there?" This had been a thrilling concept to me, that it was Monday in Houston and he was calling me from Tuesday. Now this filled me with sadness—everyone I loved was part of the past.

I allowed myself ten minutes and then it was done. To calm myself further, I repeated Eleanor's quote like a mantra: *"You must do the thing you think you cannot do."* Marie and Henri, bless them, lingered over dinner to give me time to pull myself together. When they returned, I was smiling and we made polite conversation before bed.

The acclimation day must have worked. I had no shortness of breath as Dismas and I hiked across the reddish-gray sand of the alpine desert to Kibo Hut the next morning. Henri and Marie were somewhere farther up the trail moving at a breakneck pace. The landscape was decidedly Martian, completely devoid of trees. Just a desolate smattering of rocks, no higher than a foot, and Kili, looming before us. Though we were hours away I could already make out our trail snaking steeply up the mountain, the one we would follow to the summit later that night. It was lighter than the rest of the mountain from years of scuffing by hundreds of thousands of boots. The temperature was growing colder, and I was now wearing several layers, my heaviest coat, and gloves.

Something was coming toward us now. It was a man, one of the summit hikers, being pushed down the path in a three-wheel wheelbarrow stretcher, ensconced like a pharaoh in his sleeping bag, his face barely visible. As they rolled past us Dismas exchanged a few words in Swahili with one of the porters steering the wheelbarrow.

"He had great pain in head," Dismas translated for me. "Began losing balance and couldn't walk." I nodded and tried not to think about the statistic that one hiker each month died of cerebral edema. High-altitude cerebral edema was often fatal because it required immediate medical attention and usually struck at the highest altitudes when you were already several days into your climb. The descent to the nearest medical facility was a long and precarious one.

I changed the subject. "Are the glaciers really disappearing?"

"Yes. In twenty years? No more," he said solemnly. He pointed to the base of the mountain. "See that white roof? That is Kibo Hut. Glaciers used to stretch down to there." I knew this was the dry season, but I was shocked at the lack of snow on the mountain. It was completely brown except for one sad little glacier off to the left, perched on the mountain like a too small toupee.

When we stopped for lunch, the paraplegic was there, waiting patiently in his chair while his friends took turns feeding him. Seven hours after leaving Horombo, we reached 15,000 feet and arrived at Kibo Hut, the last campground before the summit. Unlike the other camps, which had huts, Kibo consisted of one communal building, a primitive stone structure with a tin roof, perpetually glinting in the sunlight. Inside there was one long stone corridor lined with dorm rooms full of bunk beds. The hallway led to a small dining room at the back of the building with picnic tables where hikers would eat a light dinner at 5:00 P.M. and then, after a few hours of sleep, a midnight snack before leaving for the summit. There was electricity powered by solar panels but no running water and no heat. Because there was no sun, it felt colder inside. We started adding clothing immediately. When Henri took the temperature inside our cabin, it read zero degrees Celsius.

We nibbled at a dinner of soup and porridge, which I thought was only eaten by Goldilocks, three disproportionately sized bears, and Oliver Twist. And let me tell you, I don't know what Oliver was thinking when he asked for that second helping. After dinner I hurried back into Kibo Hut. I held up my hands in front of me. The high altitude was causing my body to swell, especially my face. When I'd used my tiny travel mirror to apply sunscreen earlier, I'd felt like I was looking at my reflection on the back of a spoon. I thought of that old Steve Martin joke. "I like a woman with a head on her shoulders. I hate necks."

They came for us at 11:30 P.M. and plied us with tea and cookies. I put on every piece of clothing I'd brought except for the T-shirts and shorts. I was wearing five layers on top, long johns, fleece pants, and wind pants. I dropped a few air-activated hand-warmer sacks into the knit mittens I'd borrowed from Jessica. Each of us had our own porter to walk with us up the mountain. He would carry our pack, stocked with water, Gatorade, and snacks to keep our energy up along the route.

It was going to take us approximately five hours to reach the crest of the volcano, known as Gilman's Point. We'd rest there briefly, then walk for another hour along the rim of the volcano, rising another 688 feet in elevation to reach the summit of Kilimanjaro, called Uhuru Peak, at 19,340 feet. Marie and Henri would walk with the assistant guide and the guy who'd been acting as our waiter all week. Dismas would escort me up the mountain. The rest of the porters would head back down to the Horomo Hut, where we'd meet them this afternoon. All of the groups we'd been bunking with at the various campgrounds were hiking to the summit at the same time. We gave silent nods of good luck as we walked en masse from the hut to the trail.

"Why does everyone do the summit hike at night?" I asked Dismas, over the sound of shoes crunching softly across dirt.

"Best for tourists. Sunrise is best weather and views." He paused, trying to decide whether to tell me this next part. "Also, it is so hikers cannot see steepness of mountain and how far they are from top. If they knew this, most would give up, turn back."

At first the dirt was solid. As the trail got steeper, there was more sulfurous ash mixed in with the small rocks. It was not unlike the cremains, actually. Then the rocks disappeared, and the trail became downright fluffy. We were trying to walk uphill in an ashtray. With every step, I stabbed the toes of my boots into the mountain to keep from sliding backward. Marie and Henri quickly disappeared before me, but Dismas and I were making good time, passing other groups. It was a meditative process. For hours I trained my eyes on the circle of light on the ground, watching Dismas's heels bob up and down. I was reminded of that old saying: *How do you eat an elephant? One bite at a time.*

Every now and then I glanced back and was startled by the sharp decline of headlamps dotting the trail behind me.

"Do not look around, Miss Noelley," Dismas cautioned when he saw my spotlight whirling away from the ground. "Just look at ground in front of you." Other than a forceful push, I couldn't imagine anything that would make me go down the mountain now. To throw away so many miles of hiking? Besides, the only thing worse than climbing up this mountain would be climbing down backward in the dark. As with the rickety trapeze ladder from a year ago, going up was the lesser of two evils.

Because this altitude offered only 50 percent of the oxygen at sea level, my job was to keep as much oxygen flowing to my brain as possible. My frozen nose hairs stood firm as stalagmites and stalactites, slicing the walls of my nostrils when they contracted with breath. I breathed through my mouth instead, taking vulgar goldfish gulps. Every few minutes I blew my nose and my face towel came away full of bloody scabs mixed in with the frozen boogers. The area beneath my nose was ravaged from excessive blowing, a Hitler mustache of pink raw skin. Hikers could be heard vomiting on the trail behind me.

Once in a while, Dismas asked, "You okay, Miss Noelley?"

"Yup! I'm fine." I still had no symptoms of altitude sickness, not even a headache. My legs were holding up beautifully. I was aware that it was cold, but as long as I kept moving it was not uncomfortable.

"This very surprising from girl who never hike before." He shook his head in amazement. "Most people. They very very tired by now."

Four hours into the hike the terrain turned to boulders. This was a real game-changer. Some of the rocks were bigger than me. The incline was such that I was reduced to scrambling over them on all fours. Now that I was using my arm muscles and back muscles, my body began to fatigue. I couldn't seem to get enough air. I had to rest every ten minutes. I was slightly ahead of the church group and was determined not to be overtaken by them. I didn't want to get stuck behind a human traffic jam. During my rests, I sat on a boulder and watched their line of headlamps making their way toward me like a string of belligerent Christmas lights. When they started getting too close, I reluctantly heaved myself to my feet and told Dismas, "Okay, I'm ready to go."

It's hard to keep perspective when you're on a mountain because that's the one thing you don't have—perspective. I was too close to the mountain to make sense of it. I had no idea where I was in relation to the top. I'd think I was about to scale the final crest, but when I'd reach the top, there was another crest—never before seen! I'd clamber over the next "peak" only to find another, higher ridge.

"The mountain keeps making more mountains!" I wheezed to Dismas during one of our rests. "How much farther to Gilman's Point?" I was no longer concentrating on the ground directly in front of me, but obsessing about the goal. I jammed a chocolate bar, now chalky from the cold, into my mouth.

"We are eighty-five percent of the way there, Miss Noelley."

"Eighty-five percent?!" I exclaimed, and a few clumps of chocolate fell to the dirt. "Are you effing kidding me? I thought we were, like, ninety-five percent!"

Now we were moving again. I was suffocating. Can't. Breathe. I fumbled with the snap buttons of my jacket. In one movement I tore open the front. "Get off me!" I shrieked at the jacket, as if it were an animal that had leaped onto my back. I yanked off one glove, my fingers reaching for my throat. My pulse was racing so that the beats were almost indistinguishable. Dismas waited patiently and said nothing. He had seen this all before.

Fifteen minutes later we were standing alone in front of a plaque announcing that we were at Gilman's Point. Dismas stole away to pee. I gazed at the lights zigzagging up the mountain, each dot representing a different hiker.

"Whooooooo-hoooooo!" I whooped into the darkness.

"You got that right, bitch!" someone—presumably not a member of the church group—hollered back.

I tried to take a sip of the water, but it was frozen. I sampled the Gatorade. Also frozen. Dismas returned, and we set off to circumnavigate the ring of the volcano.

"This is very dangerous part, Noelley. *Only* step where I step."

We hugged our way around boulders to keep from plummeting into the steep crater on our right side. I knew it was there, one and a half miles across, six hundred feet deep, but couldn't see it, which was probably a good thing. The sky over the horizon was lightening, filling me with urgency. I wanted to be at Uhuru as the sun rose.

"Must go faster, must go faster," I repeated, but the words were slow and blurry, my jaw numb.

My upper lip was so raw that I'd ceased blowing my nose. The snot had been dripping down a clump of hair and frozen into a snotsicle. We walked alongside the glaciers, majestic and huge like white whales. I heard voices. I kept glancing over my shoulder at the eastern horizon. The sun. Had to beat the sun. As I entered the clearing, I heaved my fist wearily into the air. I was startled to learn that out of everyone who summited Mount Kilimanjaro that day, I finished fourth. I'd thought there had been so many more people ahead of me. The first had been a bespectacled man from New Zealand. Henri and Marie had come next, beating me by a half hour.

I thought I'd feel extremely proud of myself if I made it to the top. Instead, I felt humbled. "The only thing that will get you up that mountain is sheer iron will," Becca had said, but she was wrong. I didn't believe people flew sixteen hours and hiked for four days and turned back because they weren't determined enough. On the way up the mountain I'd seen a quadriplegic whose friends were pulling him using ropes attached to his wheelchair. Meanwhile, I'd had no altitude sickness, migraine headaches, vomiting, diarrhea, muscle pain, or even blisters. The weather had been perfect—no rain or snow. Any one of those things could've prevented me from getting to the top. Sure, willpower had *something* to do with it. For the most part, though, making it to the top was pure dumb luck, as random as getting hit by a drunk driver and ending up paralyzed from the neck down. I was no better than any of the people who hadn't made it. In a way, it almost required more courage to turn back and acknowledge your limitations. I couldn't imagine going home and having to face family and friends eagerly asking, "So did you make it to the top?" Having to face yourself and the guilt and self-punishment that comes from falling short of expectations—that's courageous. Making it to the top felt like a reminder, rather than an accomplishment. It was a reminder that in my life—just like on the mountain—I'd been incredibly lucky.

It was a damn fine view, for sure. To put it in perspective, the Empire State Building is 1,453 feet high. It would take thirteen Empire State buildings stacked on top of each other to equal the height of Kilimanjaro. The peak of Kilimanjaro was simultaneously the most impressive and least assuming location I'd ever spent time in. There was only a crude wooden sign covered in jaunty yellow writing citing the altitude and announcing itself as the tallest freestanding mountain in the world. It was as if the sign knew it couldn't compete with the view, so it didn't even try. Here I was looking at these beautiful glaciers that probably wouldn't be here in twenty years. I could actually see the curve of the earth. It suddenly registered that between this moment and skydiving, I'd seen the curve of the earth twice this year.

There was a sharp snapping sound, and I looked over to see Dismas lifting a blue-and-silver can to his mouth.

"Are you seriously drinking a Red Bull right now?" I laughed, staring in disbelief at the supercaffeinated energy drink. That it was the only liquid that didn't freeze at negative ten degrees should concern us all.

My camera batteries had died as soon as I got to the summit. Thank goodness Becca had warned me to bring extras. Dismas took pictures of me standing in front of the wooden marker holding the handmade cardboard signs I'd colored. I couldn't wait to surprise my parents and Matt with the photos when I got home. Marie, Henri, and I smiled at one another and posed for a group picture, but we did not hug. I felt detached from them now. I'd done this without them.

Maybe it was my imagination, but over the past few days I'd gotten the impression that Henri purposefully wasn't taking pictures when I was taking them, as if he didn't want to concede that I'd chosen a good shot. I'd raise my camera and he'd immediately lower his. When I glanced over now, his camera was hanging at his side.

Incredulous, I asked him, "Aren't you going to take a picture of the glaciers?"

"No," he said stubbornly.

I shrugged and turned away from him and gasped. On the other side of the mountain—opposite of the sunset side—the layer of clouds below was so seamless and stretched so far that it took me a few moments to realize it wasn't snow. The day was clear and the dark shadow of Mount Kilimanjaro cast perfectly onto the radiant white. So many things had to come together for this moment to happen—I had to be here exactly at sunrise, it had to be a clear, non-snowing day, and the clouds had to be thick enough to form a white canvas for the shadow.

"It's amazing!" I cried, lifting my camera. "You have to get a picture of this!"

Henri glanced at it dismissively and walked away.

Now that we were no longer moving, the cold was penetrating. Every moment my hand was out of the mittens taking pictures of the sunrise was torturous, but I wanted to capture everything. The colors were constantly developing, luminous gold erupting into searing coppers, then giving way to aching purples and brilliant blues. Each time I decided I'd found my favorite color of the sunrise, it melted into another more spectacular shade than the last. It was, quite simply, the most magnificently beautiful thing I'd ever beheld.

We'd been cautioned about "mountain madness," where people became delirious due to the reduced oxygen. So far none of us were exhibiting any bizarre behavior, but Dismas wasn't taking any chances. After twenty minutes, Dismas said, "We go now. Not good to stay in my main office too long."

The descent took three hours. Slogging down the mountain in reverse was a rare opportunity to revisit one's accomplishments, to see how far you'd come. Dismas had been right. If I'd known what was before me, I might have turned back. In the boulder section, we stepped from rock to rock, praying they didn't give way and start an avalanche. When we came to the volcanic ash zone, it stretched before us steeply and endlessly.

I looked at Dismas wearily, "Can't I just curl up in the middle of an inner tube and roll myself down the mountain?"

Without the presence of footholds, it was too steep to tread upon. Instead we "skied" down the mountain. This was accomplished using the painful combination of leaning backward and bending one's knees as we slid our feet through the ash. I can say with confidence that my knees will never be the same. We trudged into Kibo Hut dusted in sulfur. Dismas allowed us only an hour to sleep. Removing just my boots, I gingerly slid into my sleeping bag and lay awake with my eyes closed.

Then we were back on the trail. Marie and Henri sallied forth, which was just as well.

Dismas and the assistant guide accompanied me instead. They hung back, chatting in Swahili. They were like parents trying to enjoy their adult conversation, but keeping an eye on their toddler tottering out in front of them. It occurred to me that maybe they all disliked me and didn't want to walk with me, but I couldn't muster the energy to care. I had no idea how I was going to walk seven and a half more miles. It seemed impossible.

"In some ways the hike down is the hardest part because you have nothing to work toward anymore," Becca had said.

As a general rule, uphill is hard on the muscles, downhill is hard on the bones and skin. I walked stiff-gaited, the Tin Man searching Oz for his lost oil can so he could grease his knee joints. Blisters rose on my toes where they were bumping against the front of my boots. I unzipped a side pocket on my backpack and pulled out something white and plastic. I'd avoided using my iPod so far because I'd wanted to experience Kilimanjaro with all five senses. I wanted to be fully present. But sometimes music was the only thing that could get you through the pain. For six hours I walked and stumbled down the slanted trail, working my way through my playlist. When my iPod died a few hours later, I was jealous. At least it got to stop. I recited poems to keep my woozy mind occupied. *I grow old . . . I grow old . . . I shall wear the bottoms of my trousers rolled!* When I reached the moorlands, which I'd come to think of as Seussland, I tried to remember the words to *Oh, the Places You'll Go!* I'd read the poem so many times when I made that video to get into Yale that I'd once known it by heart.

"You'll join the high fliers who soar to high heights," I murmured to myself.

I tripped on a rock and stumbled a few steps. "Poly-poly, Miss Noelley!" Dismas called out behind me. Suddenly I noticed that I felt a little high. Was this altitude-induced brain damage? Did brain-damaged people say things like "induced"? I decided that if I could remember the rest of the poem, it would prove I wasn't brain damaged.

> *On and on you will hike.*
> *And I know you'll hike far*
> *and face up to your problems*
> *whatever they are.*

"KID, YOU'LL MOVE MOUNTAINS!" I babbled out loud.

> *You're off to Great Places!*
> *Today is your day!*
> *Your mountain is waiting.*
> *So . . . get on your way!*

"What you say, Miss Noelley?" Dismas called.

Fifteen minutes later I was staggering among the black triangular cabins of Horombo. I bought a celebratory Coke, but knowing we were going to bed in a few hours, I saved it until morning. As we were getting ready for bed, I blew my nose for the five hundredth time. "Come on, Noelle, the porters have enough to carry," Marie joked of my sodden face towel, and I contemplated killing her in her sleep. That night, however, I enjoyed the best night of sleep I'd had on the mountain. It took an hour and a half to doze off, but that was a long way from lying awake for six hours, wondering how hard I'd have to hit my head to lose consciousness but not do any real damage. When I woke up the next morning, I truly believed, for the first time in years, that I'd be able to break my ten-year dependency on sleeping pills. It was hard not to lean on the crutch when it was always in hand. But here I'd have been putting my life at risk in a very direct way if I'd taken them. For me it had taken being in a situation where it wasn't an option to learn I didn't need them.

Preserved by the meat-locker air of our cabin, the Coke was exquisitely cold when I cracked it open at breakfast. I gulped it down greedily. It sparkled over my tongue and left a satisfying, crackling burn in my throat. It was, without exaggeration, the best thing I'd ever tasted.

Though more than twice the length at twelve and a half miles, today's hike would be far less brutal than yesterday's trek from Kibo to Horombo. But because of yesterday's downhill "skiing," my knees were screaming in protest. I moved in a slow zombie lurch. Marie and Henri charged down before me. What the hell were their overachieving asses trying to prove? I grumbled to Dismas, "They realize we're all sharing the same van back to Arusha and they are just going to have to wait for me in the parking lot, right?"

He grinned. "Poly-poly, Noelle."

We stopped for lunch at the Mandara Huts, where I bought another Coke. It was just as crisp and transporting as the one from that morning. Marie and Henri had already left, so I ate alone. To pass time I took out my digital camera and scanned through my summit photos. As I was clicking through the cardboard sign photos, a nervous tingle began to mount. Oh no. My heart was pounding. When I got to the last photo, my heart sank completely. I hadn't taken a photo with the I ♥ MATT sign. There were countless photos of me holding my mom's and dad's signs and even that stupid I'M HIGH sign. I'd thought I'd gotten all of them. How could I have *forgotten the Matt sign?* A once-in-a-lifetime opportunity and I'd missed it.

I continued to berate myself as we hiked through the rain forest, stepping over logs and admiring the streams with their mini-waterfalls. I considered contacting a friend of mine who was a photo editor at a magazine I'd once worked at. He could take a picture of me holding the Matt sign, then superimpose the necessary element into one of the other summit photos. Or maybe I could Photoshop one of the other signs to read I ♥ MATT? Then, a horrifying thought. Was my forgetting the Matt sign a sign that we were not supposed to be together? Or had I just been distracted by the view and my accomplishment, not to mention exhausted, after all I'd been through? I wrestled with these questions for hours but ultimately decided to just let it go. Sometimes a sign is just a sign.

Finally I emerged, sweaty and pink-faced, from the rain forest. In the parking lot, Marie, Henri, and I pooled our money and tipped the porters and guides for taking care of us all week. The tips took nearly all of my $300. Still, I slipped Dismas an extra $20, which didn't feel like enough, even though the average wage in Tanzania was less than $1 a day.

We stood in line to sign our names in a book logging everyone who reached the top of Mount Kilimanjaro. When it was my turn, I bellied up to the counter and leaned over the book. There were so many names! The rows were neatly cordoned off so no signature could be bigger than another. It was completely without ceremony, like a very long roll call. When this book filled up, it would be replaced with a fresh one for people to sign. My book would be taken away—who knows where they go?—and join the books that had come before it. The thought of this comforted me, for some reason. The female clerk pointed at the next blank space on the page. It was just a small line. But it was all that I needed. With a smile, I picked up the pen to sign my name. I am here.

Epilogue

~

About the only value the story of my life may
have is to show that one can, even without any
particular gifts, overcome obstacles that seem
insurmountable if one is willing to face the fact that
they must be overcome; that, in spite of timidity
and fear, in spite of a lack of special talents, one can
find a way to live widely and fully.

—Eleanor Roosevelt

On the morning of my thirtieth birthday, I woke up to find
that—completely independent of each other—Jessica and Chris
had sent me the same birthday card. It featured a cartoon of a
bikini-clad woman on water skis. In the speech bubble over her head,
it said, *I'm so glad your birthday will bring together all of your friends at a
time when my tan is fully realized.* Chris had added his own personal
message in the space beneath: "This is exactly what I look like in a
two-piece, actually. Have a good day and remember: You only turn 23
once. xoxo, Chris"

"So how does it feel?" Bill asked when he called to wish me a happy
birthday.

"I think I saw my first sign of crow's-feet in the mirror this morning," I said ruefully.

"I have crow's-feet on my balls now. But don't worry, that doesn't happen until thirty-three. And only if you have balls."

In the three weeks I'd been back from Africa, I'd mostly been conquering everyday fears. Kilimanjaro really felt like the culmination of the project and I'd been winding down ever since. I was sad that it was ending, of course, but it was time.

So many things had changed over the course of the year. I was still making milkshakes, but Becca was moving to Boston to start medical school. Josh and Monique moved to Berkeley and were planning their wedding. Cub and Chris moved in together and were talking about getting married. Jessica had a boyfriend with whom she was blissfully happy and went on all sorts of backpacking adventures. My little sister took a break from swimming so she could relax and enjoy life a little more, but said she might go back to it eventually. Lorena, my old coworker who'd called to tell me about the layoff, was so inspired by my project that she quit her job and moved to Australia for a year. Many things had changed. Except Bill. He was exactly the same.

"It's been a hell of a year for you, Noelle!" Dr. Bob had said in our session the other day. "Eleanor would be so proud of all you've accomplished!"

"Well, thanks," I said, feeling a little sheepish. "Eleanor changed the world. I just changed myself," I added.

He leaned back in his chair. "I don't think you needed to *change* yourself. I think you needed to *discover* yourself."

What I discovered was that, in taking on tangible challenges, I'd grown into someone who could handle the intangibles. That life was not about attaining; it was about letting go. When I looked back, nothing was ever as bad as I thought it would be. In fact, it was usually better than I could have imagined. I learned that we should take each moment both more and less seriously because everything passes. The joyful moments are just as fleeting as the terrible ones.

For my thirtieth birthday, I decided to mark the occasion in a more traditional way . . . just a party—no swinging from things, no signing of contracts "in the event of accidental death or dismemberment." At first I'd said I didn't want a big fuss. I didn't have the energy to plan anything else this year. But Matt had insisted that entering a new decade demanded a big celebration. He'd rented out a bar and drafted a bunch of my friends to help plan the festivities. There were rumors of a slide show.

Before the party, Matt took me to dinner and we went back to my apartment to loll on the couch and have a few glasses of red wine. Across the room, my parakeets were having one of their domestic arguments involving loud, indignant squawking. Nothing and everything was different.

"Happy birthday, sweetheart," he said, clinking his glass against mine. "Ready for your present?"

I nodded eagerly, but inside I was uneasy. Matt's gifts were always a bit of a wild card. Two birthdays ago he'd bought me a beautiful jade necklace. The year after that he'd given me a handheld Oriental fan and a miniature porcelain tea set, which he'd suggested I put in my parakeets' cage for decoration. I got ready to deploy my best fake smile.

He took my hand in his, and I felt something cold and metallic on my skin. I looked down at my wrist. It was a gorgeous sterling silver cuff.

"It's beautiful!" I breathed. I twisted my arm back and forth, admiring how the bracelet glinted in the light.

"I know you almost never wear bracelets. But I thought trying something different goes with the spirit of your project," he said. "Also, it's hard to monogram earrings." I pulled the bracelet off my wrist and peered at the inscription inside: *"You must do the thing you think you cannot do."—Eleanor Roosevelt*

Sentimental tears pooled in my eyes. How did he know? It's one of her lesser known lines. I'd never mentioned it to him.

"I looked up a bunch of her quotes," Matt explained. "At first I considered going with 'Do one thing every day that scares you,' but it feels like you're entering a new chapter of your life. The boldness of this one reminded me of you."

As I blinked back the tears, I put on a saucy grin. "So I'm bold now, am I?"

Matt shrugged. "To me," he said, "you have always been fearless."

He and Dr. Bob were right, of course. This whole time I'd thought I was trying to get back to the person I used to be, when really I was growing into the person I was always meant to be. I was relaxing into myself.

Now I was looking forward to getting out of my own head. I knew that Eleanor would approve. "There is a danger in this self-examination," she wrote. "Some people become so interested, so fascinated by this voyage of self-discovery, that they don't come out of it again. They remain completely absorbed in their self-study."

I put the bracelet back on my wrist. Focusing on myself so much this past year meant that I wasn't there much for Matt. "I'm sorry if I neglected you this year. It's been all about me." I'd once worried that I'd always feel slightly inferior because of Matt's many talents. But he supported me while I'd hogged the spotlight the past year. I'd been going on adventure after adventure, and he'd come along for the ride (well, driving mostly). It made me appreciate what we had even more.

"What do you mean? I'm always happy to support you, honey. We're a team," he said.

Something else that happened gradually over the past year was that I no longer felt like Matt was upstaging me. I'd realized that, frankly, it was up to me to make sure that he didn't. As Eleanor said, "No one can make you feel inferior without your consent." He took my hand and kissed the back of it.

I was quiet for a few seconds. It probably wasn't the right time to bring this up, but what the hell. I still had one fear left to face.

"Remember that night at the wedding in Nantucket?" I asked. "What were you going to say?"

"When?"

"When that guy at our table asked 'So are you two getting married?' "

His brow furrowed for a moment as he thought back to that night. Then he smiled. "I was going to say, 'We don't even live in the same city! First we have to live in the same city. Then we'll move in together and get engaged; then we get married.' "

Inwardly, I sighed with relief. For so long I hadn't dared to ask. I was afraid of his answer. Afraid he'd want to get married before I was ready. Afraid he'd say he never wanted to get married. Afraid he'd be evasive. Afraid whatever his answer was would upset the fantastic thing we had going. But his was the perfect answer. Everything would unfold as it should.

Matt looked around the apartment and said, "I think your furniture and my furniture will go together nicely."

"Mhhhmm." I made a contented noise. "Me too."

"So what happens now? You going to keep conquering one fear every day?"

"Actually, I don't even think I could find a fear every day. I've been struggling to find fears the last few weeks. The world isn't as scary now."

I added: "Besides, now I have to focus on finding a job!"

"And?" He laced his fingers in between mine and gave them a squeeze. "What are you going to do?"

"I don't know." The world was wide open. I smiled, just thinking about the possibilities. "I mean, I can do anything."

In 1960, two years before Eleanor died, she looked back at how much she had changed from the timid girl consumed by self-doubt. "It was not until I reached middle age that I had the courage to develop interests of my own," she said. "From that time on, though I have had many problems, though I have known grief and the loneliness that are the lot of most human beings, I have never been bored, never found the days long enough for the range of activities with which I wanted to fill them. And, having learned to stare down fear, I long ago reached the point where there is no living person whom I fear, and few challenges that I am not willing to face."

I'm not presumptuous enough to think I'll ever be as fearless as Eleanor. But she taught me that courage is a muscle. It needs to be exercised often or it'll weaken.

It will take time for me to understand all the ways that year changed me. A meaningful experience is a glass of wine. It needs to breathe and open up; it can only be fully appreciated when you return to it later. I suspect I'll return to this year many times throughout my life. With each passing birthday, the memories will blur and some may disappear entirely. But I know I'll always remember the startling sensation of diving out of the plane headfirst, the bright air pushing its way into my lungs, and the world rushing up to greet me as if to say, "Where have you been?"

In the interest of privacy, I've changed the names* or identifying characteristics of various individuals. That said, this is a work of nonfiction. The events and experiences detailed herein are true. I did, indeed, do one scary thing every day for a year, never cheating by skipping a day. Initially I planned to write about all of them, but it soon became clear the book would be a trillion pages long and likely to induce stupefaction. So I've chronicled the highlights, occasionally compressing or altering the timeline.

---

* This is really quite a challenge, by the way—coming up with fake names for forty people. Once you know someone's name it's nearly impossible to imagine them having any other name. (Also, that I was able to resist giving the rare obnoxious character an unbecoming name—like Dick—shows a great leap in personal growth.) Individuals whose names and characteristics weren't changed for purposes of discretion include Jessica, Chris, and Bill, who've never been discreet in their lives.

Acknowledgments

⌣

First and foremost, I'd like to acknowledge that this book would not exist without Eleanor Roosevelt. She inspired the idea for the project and inspired me on a daily basis as I faced my fears. Though the book is called *My Year with Eleanor*, I know that her life story will continue to inspire me for years to come. I'm forever grateful for all she has done for me.

I owe an enormous debt of gratitude to the wonderful and wise Dr. Robert Leahy for his guidance over the years and for helping me better understand fear. He is the best cognitive therapist around and a warm and generous spirit. He also has a fantastic laugh.

I'm basically in love with the Friedrich Agency; they believed in this book from the beginning and go to bat for it every day. A big thanks to the holy trinity: the unsinkable Molly Friedrich, the indispensable Lucy Carson, and the tireless foreign rights agent Paul Cirone. Their endless encouragement throughout this process has been such a blessing in my life. I don't think of them as my agents but as friends who happen to get a commission.

Thanks to the fabulous team at Ecco Press, which has been incredibly supportive of *My Year with Eleanor*. I was lucky to have as my editor the charming Lee Boudreaux, who loved Eleanor as much as I did. She took on this book with unbelievable enthusiasm and vision. Her editorial notes were always thoughtful and thorough and frequently hilarious. I'm hoping one day she'll tell me where she gets her energy. I'm also indebted to Abigail Holstein, who worked diligently alongside Lee and patiently answered my every question and email (of which there were many). Virginia "Ginny" Smith was the acquiring editor of this book but later took a job with another company. She was always a big supporter of my writing, and I'm very appreciative of the early work she put in on the manuscript.

The following individuals held my hand during my Year of Fear: the staff at Trapeze School New York and Long Island Skydiving; Robert "Boom" Powell and the rest of the highly trained staff at Air Combat USA; Avi Miller and Ofer Ben, my ridiculously fun and talented tap dance instructors at Broadway Dance Center; Good Earth Tours and the guides who helped me climb Mount Kilimanjaro, especially Dismas, the most patient man on planet Earth, who somehow resisted throwing me off the mountain when I got grouchy; Christa Rowe, who brought my rotator cuff back from the dead after I injured it in trapeze class and helped get my out-of-shape ass up that mountain.

Throughout the writing of this book, I leaned heavily on these lovely people: Ryan Fischer-Harbage and Christa Bourg, two unparalleled writing instructors whose exceptional advice brought this book to fruition. Christa is an incredible mentor and single-handedly kept me sane when I was on deadline. Lindsey James, my childhood best friend and the most well-read person I know, spent hours slogging through hundreds of pages and gave brilliant notes, all the way from Texas. The staff at the Starbucks on Ninth Street and Second Avenue— where much of this book was written—kept the caffeine coming at a steady clip.

I'd like to give big hugs to the early believers, including Allison

Yarrow, Corey Binns, Whitney Frick, Matt McCarthy, Lindsay Robertson, Joe Levy, and Neil Turitz. Rob "Not Ron" Tannenbaum, John Phillips, and Mark Lisanti are far funnier than I am and are responsible for some of my favorite lines in the book. I can't say enough thanks to my friends who showed up to the stand-up comedy show and cheered louder than everyone else's friends. Ditto to everyone who came to the trapeze recital, specifically Manish Vora, Josh Dienstag, Sara Kang, and Garrett Wheeler, who painted my name on their stomachs and started a possibly illegal cocktail party in the alley out back.

I am eternally grateful to Amanda Lerman, Lorena O'Neil, and especially Katharine Sise, who are always there for me. They listened to me freak out countless times about this book, without complaint, and talked me down from the ledge on several occasions. I absolutely adore them.

God bless my wonderful parents, Myatt and Bitsy Hancock, my brother, Jeff Hancock, and my sister, Jordan Hancock, who have been amazingly supportive. I'm very lucky to have them.

I offer my greatest gratitude to Jessica Coen, Chris Rovzar, Bill Schulz, and "Matt" (who's not really named Matt but wanted a pseudonym to keep his private life semi-private and because he was worried people would judge him for getting caught having almost sex in a bathroom at a wedding). I simply could not have written this book without them. Everything I mentioned above—people showing up to my ridiculous events and cheering loudly, listening to me freak out about the book, reading the book and offering input, coming up with better jokes than I ever could—they did all of that and more. They are my All of the Above. Without hesitation, they let me write about their lives. It's never easy to be written about, especially if you're in the media and understand the drawbacks that can come with being immortalized in print. I'm deeply grateful to have them in the book and in my life. They make everything better.

Recommended Reading

I cannot recommend highly enough the following books, which I turned to time and time again while researching *My Year with Eleanor*:

* *The Autobiography of Eleanor Roosevelt* by Eleanor Roosevelt (Cambridge: Da Capo Press, 2000)

* *You Learn by Living: Eleven Keys for a More Fulfilling Life* by Eleanor Roosevelt (New York: HarperCollins Publishers, 2011)

* *Eleanor Roosevelt: A Life of Discovery* by Russell Freedman (Solana Beach, CA: Sandpiper, 1997)

* *Our Eleanor: A Scrapbook Look at Eleanor Roosevelt's Remarkable Life* by Candace Fleming (New York: Atheneum/Anne Schwartz Books, 2005)